I07555737

ELEMENTS

OF THE

DIFFERENTIAL AND INTEGRAL

CALCULUS.

BY CHARLES DAVIES, LL. D.,

AUTHOR OF ARITHMETIC, ELEMENTARY ALGEBRA, ELEMENTARY GEOMETRY ELEMENTS OF SURVEYING, ELEMENTS OF DESCRIPTIVE GEOMETRY, ELEMENTS OF ANALYTICAL GEOMETRY, AND SHADES SHADOWS, AND PERSPECTIVE.

IMPROVED EDITION.

NEW YORK:
PUBLISHED BY A. S. BARNES & CO.
No. 51 JOHN STREET.
1852.

PREFACE.

THE Differential and Integral Calculus is justly considered the most difficult branch of the pure Mathematics.

The methods of investigation are, in general, not as obvious nor the connection between the reasoning and the results so clear and striking, as in Geometry, or in the elementary branches of analysis.

It has been the intention, however, to render the subject as plain as the nature of it would admit, but still, it cannot be mastered without patient and severe study.

This work is what its title imports, an Elementary Treatise on the Differential and Integral Calculus. It might have been much enlarged, but being intended for a text-book, it was not thought best to extend it beyond its present limits.

The works of Boucharlat and Lacroix have been freely used, although the general method of arranging the subjects is quite different from that adopted by either of those distinguished authors.

The present is a corrected, and it is hoped an improved edition. The first chapter has been entirely re-written, and some of the other parts of the work have been considerably altered.

West Point, *March*, 1843.

CONTENTS.

CHAPTER I.

CHAPTER II.

CHAPTER III.

CHAPTER IV.

CHAPTER V.

CHAPTER VI.

CHAPTER VII.

CHAPTER VIII.

INTEGRAL CALCULUS.

DIFFERENTIAL CALCULUS.

CHAPTER I.

Definitions and Introductory Remarks.

1. All the quantities which are considered in the Differential Calculus may be divided into two principal classes: *constants* and *variables*. Each constant retains the same value throughout the same investigation; but the variable quantities are subjected to certain laws of change, in consequence of which they may assume in succession, an infinite number of different values, without changing the form of the expression into which they enter.

The constant quantities are generally designated by the first letters of the alphabet, a, b, c, &c.; and the variable quantities by the final letters, x, y, z, &c.

2. If two variable quantities are so connected together that any change in the value of the one necessarily produces a change in the value of the other, they are said to be *functions of each other*.

Thus, in the expression

$$y = ax,$$

y and x are functions of each other; for, if any change be made in the value of x, a corresponding change will take place in that of y; and reciprocally.

3. When the value of one variable depends on that of another, as in the expression

$$y = ax, \qquad \text{or} \qquad y = cx^2,$$

if we attribute at pleasure any increment to one of the variables, a corresponding change will take place in the other; and hence, if one of them be supposed to increase or decrease according to any independent or *arbitrary* law, a corresponding change of the other will take place according to the law of relation which exists between them. The one to which the arbitrary increment is given, is called the *independent variable*, or simply the *variable*, and the other is called the function.

4. The relation between a function and its variable is generally expressed thus:

$$y = f(x),$$

in which f is a mere symbol, denoting that y and x are functions of each other. The expression is read, y a function of x, or y equal to a function of x.

The mutual dependence of one variable on another may also be expressed under the form

$$f(x,y) = 0$$

in which y is a function of x, and x a function of y.

5. Functions are either *increasing* or *decreasing*. An *increasing* function is that which increases when its variable increases, and decreases when its variable decreases

Thus, in the expressions

$$y = ax^2, \qquad u = (a + x)^2,$$

y and u are increasing functions of x; since, if x be increased, y and u will both increase; and if x be diminished y and u will both decrease.

A *decreasing* function is that which increases when its variable decreases, and decreases when its variable increases. Thus, in the expression

$$y = \frac{1}{x},$$

y is a decreasing function of x; for, if x is decreased y will increase; and reciprocally.

In the expression

$$y = (a - x)^2,$$

y will decrease while x increases between the limits of zero and a; but will increase with x for all values of x greater than a. Hence, y is a decreasing function of x for all values of x less than a, and an increasing function of x for all values of x greater than a.

6. Functions are either *explicit* or *implicit*. An *explicit function* is when the value of the function is directly expressed in terms of the variable on which it depends. Thus, in the expressions

$$u = bx^2, \qquad y = \sqrt{a^2 - x^2},$$

u and y are explicit functions of x.

An *implicit function* is one where the value is not directly expressed in terms of the variable. Thus, in the expressions

$$au^2 + cx^2 = bx^2, \qquad y^2 + x = a^2 - x^2,$$

u and y are *implicit* functions of x. These expressions may be written under the form

$$f(u,x) = 0, \qquad \text{and} \qquad f(y,x) = 0.$$

The relation between an implicit function and its variable may also be expressed by means of two or more equations. Thus, if we have

$$z = ay^2, \qquad \text{and} \qquad y = ax,$$

z is an implicit function of x.

These expressions may be written

$$z = f(y), \qquad \text{and} \qquad y = f(x);$$

or we may write

$$f(z,y) = 0, \qquad \text{and} \qquad f(y,x) = 0.$$

7. Functions are either *algebraic* or *transcendental.*

8. An *algebraic function* is one in which the relation between it and its variable can be expressed algebraically —that is, by addition, subtraction, multiplication, division, or the extraction of roots indicated by constant indices. Thus, in the expressions

$$u = ax^2 + cx, \qquad y = \sqrt{a^2 - x^2},$$

u and y are algebraic functions of x.

9. *Transcendental functions* are those in which the relation between the function and its variable cannot be determined by methods purely algebraic. For example:

$$u = a^x, \qquad u = \log. x, \qquad u = \sin x,$$

are transcendental functions.

When the variable enters as an exponent, it is called an *exponential function;* when it enters as a *logarithm,* it is called a logarithmic function; and when it enters as a sin., tang., cos., &c., it is called a *circular* function. Thus, in

$u = a^x$, u is an exponential function of x;

$u = \log x$, u is a logarithmic function of x;

$u = \sin x$, u is a circular function of x.

10. Although the values of the function and variable may be changed at pleasure without affecting the values of the constants with which they are connected, there is, nevertheless, a relation between them and the constants which it is important to consider.

If, in the equation

$$y = f(x),$$

a particular value be attributed either to x or y, the other will be expressed in terms of this value and the constant quantities which enter into the primitive equation. Thus, in the equation

$$y = ax + b,$$

if a particular value be attributed to x, the corresponding value of y will depend on the value assigned to x, and on a and b; or if a particular value be attributed to y, the corresponding value of x will depend on that value, and on a and b. The same will evidently be the case in the equation

$$x^2 + y^2 = R^2$$

or in any equation of the form

$$y = f(x).$$

Hence, we see that, although the *changes* which take place in the values of the function and variable are entirely independent of the constants with which they are connected, yet their *absolute* values are dependent on those constants.

11. Since the relations between the variables and constants are not affected by the changes of value which the variables may experience, it follows that, if the constants be determined for particular values of the variables, they will be known for all others.

Thus, in the equation

$$x^2 + y^2 = R^2,$$

if we make $x=0$, we have

$$y = \pm R;$$

or, if we make $y=0$, we have

$$x = \pm R.$$

12. The function y, and the variable x, may be so related to each other as to reduce to 0 at the same time Thus, in the equation

$$y^2 = 2px,$$

which may be placed under the general form

$$f(x, y) = 0, \qquad \text{or} \qquad y = f(x),$$

if we make $x = 0$, we have $y = 0$, or if we make $y = 0$, we shall have $x = 0$.

13. We have thus far supposed the function to depend on a single variable; it may however depend on several. Let us suppose, for example, that u depends for its value on x, y, and z; we express this dependence by

$$u = f(x, y, z).$$

If we make $x=0$, we have

$$u = f(y, z);$$

if we also make $y=0$, we have

$$u = f(z \;;$$

and if, in addition, we make $z = 0$, we have

$$u = a \textit{ constant},$$

which constant may itself be equal to 0.

If the function be of the form

$$u = b + ax + yx^2 + zx^2,$$

in which one of the variables is a factor of several of the terms, then, if $x = 0$, we shall have

$$u = b;$$

or, if x were a factor of all the terms, we should have, for $x = 0, \quad u = 0.$

14. Let us now examine the change which takes place in the function, in consequence of any change that may be made in the value of the variable on which it depends.

Let us take, as a first example,

$$u = ax^2, \quad (1)$$

and then suppose x to be increased by any quantity h.

Designate by u' the new value which u assumes, under this supposition, and we shall have

$$u' = a(x + h)^2,$$

or, by developing,

$$u' = ax^2 + 2axh + ah^2.$$

If we subtract the first equation from the last, we shall have

$$u' - u = 2axh + ah^2;$$

hence, if the variable x be increased by h, the function will be increased by $2axh + ah^2$.

If both members of the last equation be divided by h, we shall have

$$\frac{u' - u}{h} = 2ax + ah,$$

which expresses the ratio of the increment of the variable to that of the function.

Let us take, as a second example,

$$u = x^3, \quad (2)$$

and suppose x to be increased by a quantity h; designating by u' the new value which u assumes under this supposition, and we shall have

$$u' = (x + h)^3,$$

and, by developing,

$$u' = x^3 + 3x^2h + 3xh^2 + h^3.$$

By transposing x^3, substituting for it its value u, and then dividing by h, we have

$$\frac{u' - u}{h} = 3x^2 + 3xh + h^2.$$

From equation (1) we have

$$\frac{u' - u}{h} = 2ax + ah;$$

and from equation (2),

$$\frac{u' - u}{h} = 3x^2 + 3xh + h^2.$$

15. Let us now observe that the numerator in the first member of each of the above equations, is the difference between the primitive function u, and the new value u', which arose from giving an increment h to the variable x, of which u is a function. Hence we see, *that the first member of each equation is equal to the increment of the function divided by the corresponding increment of the variable.*

If we examine the second members of these equations, we find a term in each which does not contain the increment h, viz.: in the first, the term $2ax$, and in the second, $3x^2$. If now, we suppose h to diminish, it is evident that the terms $2ax$, and $3x^2$, which do not contain h, will remain unchanged, while all the terms which contain h will diminish. Hence, the ratio

$$\frac{u' - u}{h},$$

in either equation, will change with h, so long as h remains in the second member of the equation; but of all the ratios which can subsist between

$$\frac{u' - u}{h},$$

is there one which does not depend on the value of h? We have seen that as h diminishes, the ratio in the first equation approaches to $2ax$, and in the second to $3x^2$; hence, $2ax$ and $3x^2$, are the *limits* towards which the ratios approach in proportion a h is diminished; and hence, each expresses that particular ratio which is independent of the value of h. This ratio is called *the limiting ratio of the increment of the variable to the corresponding increment of the function.*

16. We are now to explain the notation by means of which this limiting ratio is to be expressed. For this purpose let us resume the equation

$$\frac{u'-u}{h} = 2ax + ah,$$

and represent by dx the last value of h, that is, *the value of* h, *which cannot be diminished, according to the law of change to which* h *or* x *is subjected,* without becoming 0 and let us also represent by du the corresponding difference between u' and u; we then have

$$\frac{du}{dx} = 2ax.$$

The letter d is used merely as a characteristic, and the expressions du, dx, are read, *differential* of u, *differential* of x.

It may be difficult to understand why the value which h assumes in passing to the limiting ratio, is represented by dx in the first member, and made equal to 0 in the second. We have represented by dx the *last* value of h, and this value forms no appreciable part of h or x. For, if it did, it might be diminished without becoming 0, and therefore would not be the *last* value of h. By designating this last value by dx, we preserve a trace of the letter x, and express at the same time the last change which takes place in h, as it becomes equal to 0. For a like reason the last difference between u' and u is designated by du.

The limiting ratio in equation (2) is

$$\frac{du}{dx} = 3x^2.$$

The limiting ratio of the increment of the variable to that of the function, which has been found in the preceding equations, is called *the differential coefficient of* u *regarded as a function of* x.

17. Let us take, as another example, the function

$$u = ax^4: \quad (3)$$

if we give to x an increment h, we shall have

$$u' = ax^4 + 4ax^3h + 6ax^2h^2 + 4axh^3 + ah^4, \quad \text{and}$$

$$\frac{u' - u}{h} = 4ax^3 + 6ax^2h + 4ah^2 + ah^3,$$

and, by taking the limiting ratio, we have for the differential coefficient,

$$\frac{du}{dx} = 4ax^3.$$

18. If it were required to find the differential of the function u, after we had formed its differential coefficient, it could be done by simply *muliplying the differential coefficient by the differential of the variable;* thus, from equation (1) we should have

$$du = 2axdx;$$

from the second,

$$du = 3x^2dx;$$

and from the third,

$$du = 4ax^3dx.$$

The differential of each function may also be written under the following form:

Eq. 1, $$\frac{du}{dx}dx = 2axdx;$$

Eq. 2, $$\frac{du}{dx}dx = 3x^2dx;$$

Eq. 3, $$\frac{du}{dx}dx = 4ax^3dx;$$

which, indeed, is nothing more than finding the differential of the function *by multiplying the differential coefficient expressed in the first member of the equation, by the differential of the variable.*

19. Let us now examine each of the three equations which we have considered, and observe the *form* of the expression for the difference between the two states of the function u.

From the first equation we have

$$u' - u = 2axh + ah^2;$$

from the second,

$$u' - u = 3x^2h + 3xh^2 + h^3;$$

and from the third,

$$u' - u = 4ax^3h + 6ax^2h^2 + 4axh^3 + ah^4.$$

We see in each of the expressions for the difference between the two states of the function u, that the first term of the difference contains the first power of the increment h, and that the *coefficient* of this term is the *differential coefficient of the function* u, or the limiting ratio of the increment of the function to that of the variable. This differential coefficient is, in general, a function of x.

If, now, in either of the expressions, we represent the differential coefficient, or limiting ratio, by P, and all the following terms of the difference by $P'h^2$ (in which P' will in general be a function of h), the difference may be written under the form

$$u' - u = Ph + P'h^2;$$

and we shall assume that what has been proved in regard to the three forms of the function u which we have considered, is equally true for all other forms. This form, for the difference between the two states of the function, is important, and should be carefully remembered. If, then, we have a function of the form

$$u = f(x),$$

and give to x an increment h, we shall have

$$u' - u = Ph + P'h^2.$$

If, now, we wish the ratio of the increments, we have

$$\frac{u' - u}{h} = P + P'h,$$

and, passing to the limiting ratio,

$$\frac{du}{dx} = P;$$

and, if we wish the differential of the function u, we have

$$du = Pdx,$$

or

$$\frac{du}{dx}dx = Pdx.$$

If we represent the increment of the variable by k, and

the differential coefficient by Q, the difference would be represented by

$$u' - u = Qk + Q'k^2$$

and
$$\frac{du}{dx}dx = Qdx.$$

We may conclude from the above, that if we have the difference between two states of a function, as

$$u' - u = Ph + P'h^2,$$

that we can immediately pass to the differential of u, *by writing* du *for* u′—u, *substituting* dx *for* h *in the second member, and omitting the terms which contain* h².

20. If two functions, u and v, dependent on the same variable, are *equal* to each other, for all possible values which may be attributed to that variable, the differentials of those functions will also be equal.

For, suppose x to be the independent variable. We shall then have (Art. 15),

$$u' - u = Ph + P'h^2,$$
$$v' - v = Qh + Q'h^2,$$

in which Q is the differential coefficient of v, regarded as a function of x.

But, since u' and v' are, by hypothesis, equal to each other, as well as u and v, we have

$$Ph + P'h^2 = Qh + Q'h^2,$$

or, by dividing by h and passing to the limiting ratio,

$$P = Q,$$

hence,
$$\frac{du}{dx} = \frac{dv}{dx},$$

and $$\frac{du}{dx}dx = \frac{dv}{dx}dx,$$

that is, the differential of u is equal to the differential of v.

21. The reverse of the above proposition is not generally true; that is, *if two differentials are equal to each other, we are not at liberty to conclude that the functions from which they were derived are also equal.*

For, let

$$u = v \pm A,$$

in which A is a constant, and u and v both functions of x. Giving to x an increment h, we shall have

$$u' = v' \pm A,$$

from which subtract the primitive equation, and we obtain

$$u' - u = v' - v,$$

and, by substituting for the difference between the two states of the function, we have

$$Ph + P'h^2 = Qh + Q'h^2$$

Dividing by h, and passing to the limiting ratio, we obtain

$$P = Q: \text{ that is } \frac{du}{dx} = \frac{dv}{dx};$$

hence, $$\frac{du}{dx}dx = \frac{dv}{dx}dx;$$

or, what is the same thing, by merely changing the form,

$$du = dv.$$

Here we see that although v may be greater or less than u by the constant quantity A, still its differential will always be equal to that of u.

Hence, also, we conclude *that every constant quantity, connected with the variable by the sign plus or minus, will disappear in the differentiation.*

The reason of this is apparent; for, as a constant admits of no increase or decrease, there is no ultimate or last difference between two of its values; and this *ultimate or last difference* is the differential of a variable function. Hence the differential of a constant quantity is equal to 0.

22. If we have a function of the form

$$u = Av,$$

in which u and v are both functions of x, and give to x an increment h, we shall have

$$u' - u = A(v' - v),$$

or $$Ph + P'h^2 = A(Qh + Q'h^2);$$

and, by dividing by h, and passing to the limiting ratio, we have

$$P = AQ,$$

or $$Pdx = AQdx.$$

But $$Pdx = du, \quad \text{and} \quad Qdx = dv,$$

hence, $$du = Adv;$$

that is, *the differential of the product of a variable quantity by a constant, is equal to the differential of the variable multiplied by the constant.*

CHAPTER II.

Differentiation of Algebraic Functions—Successive Differentials—Taylor's and Maclaurin's Theorems.

23. Algebraic functions are those which involve the sum or difference, the product or quotient, the roots or powers, of the variables. They may be divided into two classes, real and imaginary.

24. Let it be required to find the differential of the function.

$$u = ax.$$

If we give to x an increment h, and designate the second state of the function by u', we shall have

$$u' = ax + ah = u + ah,$$

$$\frac{u' - u}{h} = a:$$

hence, $\quad du = adx, \quad$ or $\quad \frac{du}{dx}dx = adx.$

25. As a second example, let us take the function

$$u = ax^2.$$

If we give to x an increment h, we have

$$u' = ax^2 + 2ahx + ah^2,$$

$$\frac{u' - u}{h} = 2ax + ah:$$

hence, $$du = 2ax\,dx.$$

26. For a third example, take the function

$$u = ax^3:$$

giving to x an increment h, we have

$$\frac{u' - u}{h} = 3ax^2 + 3axh + ah^2,$$

or passing to the limit

$$\frac{du}{dx} = 3ax^2; \quad \text{hence,} \quad du = 3ax^2\,dx.$$

27. Let us now suppose the function u to be composed of several variable terms: that is, of the form

$$u = y + z - w = f(x),$$

in which y, z, and w, are functions of x.

If we give to x an increment h, we shall have

$$u' - u = (y' - y) + (z' - z) - (w' - w):$$

hence, (Art. 19),

$$u' - u = (Ph + P'h^2) + (Qh + Q'h^2) - (Lh + L'h^2),$$

or, $$\frac{u' - u}{h} = (P + P'h) + (Q + Q'h) - (L + L'h)$$

or by passing to the limit

$$\frac{du}{dx} = P + Q - L,$$

and multiplying both members by dx, we have

$$\frac{du}{dx}dx = Pdx + Qdx - Ldx.$$

But since P, Q, and L, are the differential coefficients of y, z, and w, regarded as functions of x, it follows (Art. 18) that, *the differential of the sum, or difference of any number of functions, dependent on the same variable, is equal to the sum or difference of their differentials taken separately.*

28. Let us now determine the differential of the product of two variable functions.

If we designate the functions by u and v, and suppose them to depend on a variable x, we shall have

$$u' = u + Ph + P'h^2,$$

$$v' = v + Qh + Q'h^2,$$

and by multiplying

$$u'v' = (u + Ph + P'h^2)(v + Qh + Q'h^2)$$

$$= uv + vPh + uQh + PQh^2 + \&c;$$

hence

$$\frac{u'v' - uv}{h} = vP + uQ + \text{terms containing } h, h^2, \& h^3.$$

If now, we pass to the limiting ratio, we have

$$\frac{d(uv)}{dx} = vP + uQ;$$

therefore, $d(uv) = vPdx + uQdx = vdu + udv$.

Hence, *the differential of the product of two functions dependent on the same variable, is equal to the sum of the products which arise by multiplying each by the differ ential of the other.*

If we divide by uv, we have

$$\frac{d(uv)}{uv}=\frac{du}{u}+\frac{dv}{v};$$

that is, *the differential of the product of two functions, divided by the product, is equal to the sum of the quotients which arise, by dividing each differential by its function.*

29. We can easily determine from the last formula, the differential of the product of any number of functions.

For this purpose, put $v=ts$, then

$$\frac{dv}{v}=\frac{d(ts)}{ts}=\frac{dt}{t}+\frac{ds}{s},$$

and by substituting for v in the last equation, we have

$$\frac{d(uts)}{uts}=\frac{du}{u}+\frac{dt}{t}+\frac{ds}{s};$$

and in a similar manner, we should find

$$\frac{d(utsr\ldots.)}{utsr\ldots.}=\frac{du}{u}+\frac{dt}{t}+\frac{ds}{s}+\frac{dr}{r}\ldots.\ \&c.$$

If in the equation

$$\frac{d(uts)}{uts}=\frac{du}{u}+\frac{dt}{t}+\frac{ds}{s},$$

we multiply by the denominator of the first member, we shall have

$$d(uts)=tsdu+usdt+utds;$$

and hence, *the differential of the product of any number of functions, is equal to the sum of the products which arise by multiplying the differential of each function by the product of all the others.*

30. To obtain the differential of any fraction, as $\frac{u}{v}$ we make

$$\frac{u}{v}=t, \quad \text{and hence} \quad u=tv.$$

Differentiating both members, we find

$$du=vdt+tdv;$$

finding the value of dt, and substituting for t its value $\frac{u}{v}$, we obtain

$$dt=\frac{du}{v}-\frac{udv}{v^2},$$

or by reducing to a common denominator

$$dt=\frac{vdu-udv}{v^2};$$

hence, *the differential of a fraction is equal to the denominator into the differential of the numerator, minus the numerator into the differential of the denominator, divided by the square of the denominator.*

31. If the numerator u is constant in the fraction $t=\frac{u}{v}$, its differential will be 0 (Art. 21), and we shall have

$$dt=-\frac{udv}{v^2}, \quad \text{or} \quad \frac{dt}{dv}=-\frac{u}{v^2}.$$

When u is constant, t is a decreasing function of v (Art. 5), and the differential coefficient of t is negative.

This is only a particular case of a general proposition.

For, let u be a decreasing function of x. Then, if we give to x any increment, as h, we have

$$u' = u + Ph + P'h^2,$$

or,

$$u' - u = Ph + P'h^2.$$

But by hypothesis $u > u'$; hence, the second member is essentially negative for all values of h; and, passing to the limiting ratio,

$$\frac{du}{dx} = -P.$$

hence, *the differential coefficient of a decreasing function is negative.*

32. To find the differential of any power of a function, let us first take the function u^n, in which n is a positive and whole number. This function may be considered as composed of n factors each equal to u. Hence, (Art. 29),

$$\frac{d(u^n)}{u^n} = \frac{d(uuuu\,\ldots\,.)}{(uuuu\,\ldots\,.)} = \frac{du}{u} + \frac{du}{u} + \frac{du}{u} + \frac{du}{u} + \ldots\ldots$$

But since there are n equal factors in the first member, there will be n equal terms in the second; hence,

$$\frac{d(u^n)}{u^n} = \frac{ndu}{u};$$

therefore,

$$d(u^n) = nu^{n-1}du.$$

If n is fractional, represent it by $\frac{r}{s}$, and make

$$v = u^{\frac{r}{s}}, \qquad \text{whence,} \qquad u^r = v^s;$$

and since r and s are supposed to represent entire numbers, we shall have

$$ru^{r-1}du = sv^{s-1}dv;$$

from which we find

$$dv = \frac{ru^{r-1}}{sv^{s-1}}du = \frac{ru^{r-1}}{su^{\frac{r}{s}(s-1)}}du,$$

or by reducing

$$dv = \frac{r}{s}u^{\frac{r}{s}-1}du;$$

which is of the same form as the function

$$d(u^n) = nu^{n-1}du,$$

by substituting the exponent $\frac{r}{s}$ for n.

Finally, if n is negative, we shall have

$$u^{-n} = \frac{1}{u^n},$$

from which we have (Art. 31),

$$d(u^{-n}) = d\left(\frac{1}{u^n}\right) = \frac{-d(u^n)}{u^{2n}} = \frac{-nu^{n-1}\ du}{u^{2n}};$$

hence, by reducing

$$d(u^{-n}) = -nu^{-n-1}du.$$

Hence, *the differential of any power of a function, is equal to the exponent multiplied by the function with its primitive exponent minus unity, into the differential of the function.*

33. Having frequent occasion to differentiate radicals of the second degree, we will give a specific rule for this class of functions.

Let $\qquad v = \sqrt{u}, \qquad$ or $\qquad v = u^{\frac{1}{2}};$

then, $\qquad dv = \frac{1}{2}u^{\frac{1}{2}-1}du = \frac{1}{2}u^{-\frac{1}{2}}du = \frac{du}{2\sqrt{u}};$

that is, *the differential of a radical of the second degree, is equal to the differential of the quantity under the sign, divided by twice the radical.*

34. It has been remarked (Art. 3), that in an equation of the form

$$u = f(x),$$

we may regard u as the function, and x as the variable, or x as the function, and u as the variable. We will now show that, *the differential coefficient which is obtained by regarding* u *as a function of* x, *is equal to the reciprocal of that which is obtained by regarding* x *as a function of u.*

If we have

$$u = f(x),$$

and give to x an increment h, we have (Art. 19),

$$u' - u = Ph + P'h^2. \quad (1)$$

But, if x be expressed in u, and we have

$$x = f'(u),$$

and then give to u an increment k, we shall nave

$$x' - x = h = Qk + Q'k^2. \quad (2)$$

But $k = u' - u$. Substituting these values for $u' - u$, and h, in equation (1), and we have

$$k = PQk + \text{terms containing the higher powers of } k.$$

Dividing by k, and passing to the limiting ratio, we have

$$1 = PQ, \quad \text{or} \quad P = \frac{1}{Q}.$$

To illustrate this by an example, let

$$u = x^3, \quad \text{whence} \quad x = \sqrt[3]{u} = u^{\frac{1}{3}}.$$

Now, $$\frac{du}{dx} = 3x^2 = 3u^{\frac{2}{3}};$$

but regarding x as the function

$$\frac{dx}{du} = \frac{1}{3}u^{-\frac{2}{3}} = \frac{1}{3u^{\frac{2}{3}}}.$$

35. If we have three variables u, y, and x, which are mutually dependant on each other, the relations between them may be expressed by the equations

$$u = f(y), \quad \text{and} \quad y = f'(x).$$

If now we attribute to x an increment h, and designate by k, the corresponding increment of y, we shall have (Art. 19),

$$u' = u + Pk + P'k^2, \qquad y' = y + Qh + Q'h^2,$$

and $$\frac{u'-u}{k} = P + P'k, \qquad \frac{y'-y}{h} = Q + Q'h,$$

If we multiply these equations together, member by member, we shall have

$$\frac{u'-u}{k} \times \frac{y'-y}{h} = (P + P'k)(Q + Q'h);$$

but $k = y' - y$; hence, by dividing and passing to the limiting ratio, we have

$$\frac{du}{dx} = \frac{du}{dy} \times \frac{dy}{dx};$$

and hence, if three quantities are mutually dependant on

each other, *the differential coefficient of the first regarded as a function of the third, will be equal to the differential coefficient of the first regarded as a function of the second, multiplied by the differential coefficient of the second regarded as a function of the third.*

36. Let us take as an example

$$v = bu^3, \qquad u = ax^2,$$

we find

$$\frac{dv}{du} = 3bu^2, \qquad \frac{du}{dx} = 2ax$$

But, $$\frac{dv}{dx} = \frac{dv}{du} \times \frac{du}{dx} = 3bu^2 \times 2ax = 6abu^2x;$$

and by substituting for u^2, its value a^2x^4,

$$\frac{dv}{dx} = 6a^3bx^5, \qquad \text{and} \qquad dv = 6a^3bx^5dx.$$

EXAMPLES.

1. Find the differential of u in the expression

$$u = \sqrt{a^2 - x^2}.$$

Put $a^2 - x^2 = y$, then $u = y^{\frac{1}{2}}$, and the dependence between u and x, is expressed by means of y, and u is an implicit function of x. Differentiating, we find

$$\frac{du}{dy} = +\frac{1}{2}y^{-\frac{1}{2}} = \frac{1}{2}(a^2 - x^2)^{-\frac{1}{2}}, \qquad \text{and} \qquad \frac{dy}{dx} = -2x;$$

by multiplying the coefficients together we obtain

$$\frac{du}{dx} = -\frac{1}{2}(a^2 - x^2)^{-\frac{1}{2}}\,2x = \frac{-x}{\sqrt{a^2 - x^2}};$$

hence,

$$du = \frac{-xdx}{\sqrt{a^2 - x^2}}.$$

2. Find the differential of the function

$$u = (a + bx^n)^m.$$

Place $a + bx^n = y$: then $u = y^m$; and

$$\frac{du}{dy} = my^{m-1} = m(a + bx^n)^{m-1}.$$

$$\frac{dy}{dx} = nbx^{n-1};$$

hence,

$$\frac{du}{dx} = mnb(a + bx^n)^{m-1}x^{n-1}.$$

$$du = mnb(a + bx^n)^{m-1}x^{n-1}dx.$$

3. Find the differential of the function

$$u = x(a^2 + x^2)\sqrt{a^2 - x^2},$$

$$du = \left((a^2 + x^2)\sqrt{a^2 - x^2}\right)dx + x\sqrt{a^2 - x^2}\,d(a^2 + x^2),$$

$$+ x(a^2 + x^2)d\sqrt{a^2 - x^2},$$

in which the operations in the last two terms are only indicated. If we perform them, we find

$$d(a^2 + x^2) = d(x^2) = 2xdx,$$

$$d\sqrt{a^2 - x^2} = \frac{d(-x^2)}{2\sqrt{a^2 - x^2}} = \frac{-xdx}{\sqrt{a^2 - x^2}}.$$

Substituting these values, we find

$$du = \left[(a^2+x^2)\sqrt{a^2-x^2} + 2x^2\sqrt{a^2-x^2} - \frac{x^2(a^2+x^2)}{\sqrt{a^2-x^2}}\right]dx;$$

or, reducing to a common denominator and cancelling the like terms,

$$du = \frac{(a^4+a^2x^2-4x^4)\,dx}{\sqrt{a^2-x^2}}.$$

4. Find the differential of the function

$$u = \frac{a^2-x^2}{a^4+a^2x^2+x^4}.$$

$$du = \frac{(a^4+a^2x^2+x^4)\,d(a^2-x^2)-(a^2-x^2)\,d(a^4+a^2x^2+x^4)}{(a^4+a^2x^2+x^4)^2},$$

from which we find

$$du = \frac{-2x(2a^4+2a^2x^2-x^4)\,dx}{(a^4+a^2x^2+x^4)^2}.$$

5. Find the differential of the function

$$u = \sqrt[4]{\left(a - \frac{b}{\sqrt{x}} + \sqrt[3]{(c^2-x^2)^2}\right)^3}.$$

Make $\quad y = \dfrac{b}{\sqrt{x}}, \qquad z = \sqrt[3]{(c^2-x^2)^2},$

then we shall have

$$u = \sqrt[4]{(a-y+z)^3} = (a-y+z)^{\frac{3}{4}};$$

we therefore have (Art. 32),

$$du = \frac{3}{4}(a - y + z)^{\frac{3}{4}-1} d(a - y + z),$$

$$= \frac{3}{4}(a - y + z)^{-\frac{1}{4}}(-dy + dz),$$

$$= \frac{-3dy + 3dz}{4\sqrt[4]{a - y + z}}.$$

But from the equations above, we find

$$dy = d\left(\frac{b}{\sqrt{x}}\right) = -b\frac{d\sqrt{x}}{x} = \frac{-bdx}{2x\sqrt{x}};$$

$$dz = d(c^2 - x^2)^{\frac{2}{3}} = \frac{2}{3}(c^2 - x^2)^{\frac{2}{3}-1} d(c^2 - x^2),$$

$$= \frac{2}{3}(c^2 - x^2)^{-\frac{1}{3}} \times -2xdx = \frac{-4xdx}{3\sqrt[3]{c^2 - x^2}}.$$

Substituting these values of dy and dz, in the expression for du, we find

$$du = \left\{ \frac{\dfrac{3b}{2x\sqrt{x}} - \dfrac{4x}{\sqrt[3]{c^2 - x^2}}}{4\sqrt[4]{a - \dfrac{b}{\sqrt{x}} + \sqrt[3]{(c^2 - x^2)^2}}} \right\} dx.$$

6. $u = \dfrac{1}{x}$, $\qquad du = \dfrac{-dx}{x^2}.$

7. $u = \dfrac{1}{x^n}$, $\qquad du = \dfrac{-ndx}{x^{n+1}}.$

8. $u = \sqrt{2ax + x^2}$, $\qquad du = \dfrac{(a + x)dx}{\sqrt{2ax + x^2}}$.

9. $u = (a^2 + x^2)^3$, $\qquad du = 6(a^2 + x^2)^2 xdx$.

10. $u = a^6 + 3a^4x^2 + 3a^2x^4 + x^6$, $\qquad du = 6(a^2 + x^2)^2 xdx$.

11. $u = \dfrac{1}{\sqrt{1 - x^2}}$, $\qquad du = \dfrac{xdx}{(1 - x^2)^{\frac{3}{2}}}$.

12. $u = \dfrac{x}{x + \sqrt{1 - x^2}}$, $\qquad du = \dfrac{dx}{\sqrt{1 - x^2}\left(x + \sqrt{1 - x^2}\right)^2}$.

13. $u = \left(a + \sqrt{x}\right)^3$, $\qquad du = \dfrac{3\left(a + \sqrt{x}\right)^2 dx}{2\sqrt{x}}$.

14. $u = \left[a + \sqrt{b - \dfrac{c}{x^2}}\right]^4$, $\qquad du = \dfrac{\dfrac{4c}{x^3}\left[a + \sqrt{b^2 - \dfrac{c}{x^2}}\right]^3 dx}{\sqrt{b - \dfrac{c}{x^2}}}$.

15. $u = x^2y^2$ $\qquad du = 2x^2ydy + 2y^2xdx$.

16. $u = \sqrt{a^2 + x^2} \times \sqrt{b^2 + y^2}$, $\qquad du = \dfrac{(b^2 + y^2)xdx + (a^2 + x^2)ydy}{\sqrt{a^2 + x^2}\sqrt{b^2 + y^2}}$.

17. $u = \dfrac{x^n}{(1 + x)^n}$, $\qquad du = \dfrac{nx^{n-1}dx}{(1 + x)^{n+1}}$.

18. $u = \dfrac{1 + x^2}{1 - x^2}$, $\qquad du = \dfrac{4xdx}{(1 - x^2)^2}$.

19. $u = \dfrac{x + y}{z^3}$, $\qquad du = \dfrac{z(dx + dy) - (x + y)3dz}{z^4}$.

20. $u = \dfrac{\sqrt{1+x}+\sqrt{1-x}}{\sqrt{1+x}-\sqrt{1-x}}, \quad du = -\dfrac{dx(1+\sqrt{1-x^2})}{x^2\sqrt{1-x^2}}.$

21. Find the differential coefficient of

$$F(x) = 8x^4 - 3x^3 - 5x$$

Ans. $32x^3 - 9x^2 - 5.$

22. Find the differential coefficient of

$$F(x) = (x^3 + a)(3x^2 + b)$$

Ans. $15x^4 + 3x^2b + 6ax.$

23. Find the differential coefficient of

$$F(x) = (ax + x^2)^2,$$

Ans. $2(ax + x^2)(a + 2x).$

24. Find the differential coefficient of

$$F(x) = \frac{x}{x + \sqrt{1 - x^2}},$$

Ans. $\dfrac{1}{\sqrt{1-x^2}(1 + 2x\sqrt{1-x^2})}.$

Of Successive Differentials.

37. It has been remarked (Art. 19), that the differential coefficient is generally a function of x. It may therefore be differentiated, and x may be regarded as the independent variable. A new differential coefficient may thus be obtained, which is called the *second differential coefficient.*

38. In passing from the function u to the first differential coefficient, the exponent of x in every term in which x enters, will be changed; and hence, the relation which exists between the primitive function u and the variable x, is different from that which will exist between the first differential coefficient and x. Hence, the same change in x will occasion different degrees of change in the primitive function and in the first differential coefficient.

The second differential coefficient will, in general, be a function of x: hence, a new differential coefficient may be formed from it, which will also be a function of x; and so on, for succeeding differential coefficients.

If we designate the successive differential coefficients by

$$p, \quad q, \quad r, \quad s, \quad \&c.,$$

we shall have

$$\frac{du}{dx}=p, \qquad \frac{dp}{dx}=q, \qquad \frac{dq}{dx}=r, \quad \&c.$$

But the differential of p is obtained by differentiating its value $\frac{du}{dx}$, regarding the denominator dx as constant; we therefore have

$$d\left(\frac{du}{dx}\right)=dp, \qquad \text{or,} \qquad \frac{d^2u}{dx}=dp,$$

and by substituting for dp its value, we have

$$\frac{d^2u}{dx^2}=q.$$

The notation d^2u, indicates that the function u has been differentiated twice, and is read, *second differential of* u. The denominator dx^2 expresses *the square of the differential of* x, and not the differential of x^2. It is read, *differential square of* x, or differential of x *squared*.

If we differentiate the value of q, we have

$$d\left(\frac{d^2u}{dx^2}\right) = dq, \qquad \text{or,} \qquad \frac{d^3u}{dx^2} = dq\,;$$

hence,
$$\frac{d^3u}{dx^3} = r, \quad \&c.,$$

and in the same manner we may find

$$\frac{d^4u}{dx^4} = s,$$

The third differential coefficient $\frac{d^3u}{dx^3}$, is read, third differential of u divided by dx cubed; and the differential coefficients which succeed it, are read in a similar manner.

Hence, the successive differential coefficients are

$$\frac{du}{dx} = p, \qquad \frac{d^2u}{dx^2} = q, \qquad \frac{d^3u}{dx^3} = r, \qquad \frac{d^4u}{dx^4} = s, \quad \&c.,$$

from which we see, that each differential coefficient is deduced from the one which precedes it, in the same way that the first is deduced from the primitive function.

39. If we take a function of the form

$$u = ax^n,$$

we shall have for the first differential coefficient,

$$\frac{du}{dx} = nax^{n-1}.$$

If we now consider n, a, and dx, as constant, we shall have for the second differential coefficient

$$\frac{d^2u}{dx^2} = n(n-1)ax^{n-2},$$

and for the third,

$$\frac{d^3u}{dx^3} = n(n-1)(n-2)\,ax^{n-3},$$

and for the fourth,

$$\frac{d^4u}{dx^4} = n(n-1)(n-2)(n-3)ax^{n-4}.$$

It is plain, that when n is a positive whole number, the function

$$u = ax^n,$$

will have n differential coefficients. For, when n differentiations shall have been made, the exponent of x in the second member will be 0; hence, the nth differential coefficient will be constant, and the succeeding ones will be equal to 0. Thus,

$$\frac{d^nu}{dx^n} = n(n-1)(n-2)(n-3)\,\ldots\ldots\,a.1,$$

and, $$\frac{d^{n+1}u}{dx^{n+1}} = 0.$$

Taylor's Theorem.

40. TAYLOR'S THEOREM explains the method of developing into a series any function of the sum or difference of two variables that are independent of each other, according to the ascending powers of one of them.

41. Before giving the demonstration of this theorem, it will be necessary to prove a principle on which it depends, viz: *if we have a function of the sum or difference of two variables*

$$u = f(x \pm y),$$

the differential coefficient will be the same if we suppose x *to vary and* y *to remain constant, as when we suppose* y *to vary and* x *to remain constant.*

For, make $$x \pm y = z:$$

we shall then have

$$u = f(z)$$

and $$\frac{du}{dz} = p.$$

If we suppose y to remain constant and x to vary, we have

$$dz = dx,$$

and if we suppose x to remain constant and y to vary, we have

$$dz = dy.$$

But since the differential coefficient p is independent of dz' (Art. 15), it will have the same value whether,

$$dz = dx, \qquad \text{or,} \qquad dz = dy.$$

To illustrate this principle by a particular example, let us take

$$u = (x + y)^n.$$

If we suppose x to vary and y to remain constant, we find

$$\frac{du}{dx} = n(x + y)^{n-1},$$

and if we suppose y to vary and x to remain constant, we find

$$\frac{du}{dy} = n(x + y)^{n-1},$$

the same as under the first supposition.

42. It is evident that the

$$f(x + y),$$

must be expressed in terms of the two variables x and y, and of the constants which enter into the function.

Let us then assume

$$f(x + y) = A + By^a + Cy^b + Dy^c +, \&c.,$$

in which the terms are arranged according to the ascending powers of y, and in which A, B, C, D, &c., are independent of y, but functions of x, and dependant on all the constants which enter the primitive function. It is now required to find such values for the exponents a, b, c, &c., and the coefficients A, B, C, D, &c., as shall render the development true for all possible values which may be attributed to x and y.

In the first place, there can be no negative exponents. For, if any term were of the form

$$By^{-a},$$

it may be written

$$\frac{B}{y^a},$$

and making $y=0$, this term would become infinite, and we should have

$$f(x)=\infty,$$

which is absurd, since function of x, which is independent of y, does not necessarily become infinite when $y=0$.

The first term A, of the development, is the value which the primitive function assumes when we make $y=0$. If we designate this value by u, we shall have

$$f(x)=u.$$

If we make

$$f(x+y)=u',$$

and differentiate, under the supposition that x varies and y remains constant, we shall have

$$\frac{du'}{dx}=\frac{dA}{dx}+\frac{dB}{dx}y^a+\frac{dC}{dx}y^b+\frac{dD}{dx}y^c+ \quad \&c.:$$

and if we differentiate, regarding y as a variable and x as constant, we shall find

$$\frac{du'}{dy}=aBy^{a-1}+bCy^{b-1}+cDy^{c-1}+, \&c.:$$

But these differential coefficients are equal to each other (Art. 41); hence, the second members of the equations are equal, and since the coefficients of the series are independent of y, and the equality exists whatever be the value of y, it follows that the corresponding terms in each series will contain like powers of y, and that the coefficients of y in these terms will be equal (Alg. Art. 244). Hence,

$$a-1=0, \qquad b-1=a, \qquad c-1=b, \quad \&c.,$$

and consequently

$$a=1, \qquad b=2, \qquad c=3, \quad \&c.;$$

and comparing the coefficients, we find

$$B=\frac{dA}{dx}, \quad C=\frac{1}{2}\frac{dB}{dx}, \quad D=\frac{1}{3}\frac{dC}{dx}.$$

And since we have made

$$f(x)=A=u, \quad \text{and} \quad f(x+y)=u',$$

we shall have

$$A=u, \quad B=\frac{du}{dx}, \quad C=\frac{d^2u}{1.2\,dx^2}, \quad D=\frac{d^3u}{1.2.3\,dx^3};$$

and consequently,

$$u'=u+\frac{du}{dx}y+\frac{d^2u}{dx^2}\frac{y^2}{1.2}+\frac{d^3u}{dx^3}\frac{y^3}{1.2.3}+, \quad \&c.$$

43. This theorem gives the following development for the function

$$u'=(x+y)^n,$$

$$u=x^n, \quad \frac{du}{dx}=nx^{n-1}, \quad \frac{d^2u}{dx^2}=n(n-1)x^{n-1}, \quad \&c.:$$

hence,

$$u' = (x+y)^n = x^n + nx^{n-1}y + \frac{n(n-1)}{1.2}x^{n-2}y^2,$$

$$+\frac{n(n-1)(n-2)}{1.2.3}x^{n-3}y^3 +, \quad \&c.$$

44. The theorem of Taylor may also be applied to the development of the second state of any function of the form

$$u = f(x),$$

when x receives an arbitrary increment h, and becomes $x+h$. For, if we substitute h for y, we have

$$u' = u + \frac{du}{dx}h + \frac{d^2u}{dx^2}\frac{h^2}{1.2} + \frac{d^3u}{dx^3}\frac{h^3}{1.2.3} +, \ \&c.;$$

or
$$\frac{u'-u}{h} = \frac{du}{dx} + \frac{d^2u}{dx^2}\frac{h}{1.2} + \frac{d^3u}{dx^3}\frac{h^2}{1.2.3} +, \&c.,$$

$$= \frac{du}{dx} + h\left(\frac{du^2}{dx^2}\frac{1}{1.2} + \frac{d^3u}{dx^3}\frac{h}{1.2.3} +, \&c.\right)$$

Now, it is plain that h may be made so small that the term
$$h\left(\frac{d^2u}{dx^2}\frac{1}{1.2} + \frac{d^3u}{dx^3}\frac{h}{1.2.3} +, \&c.,\right)$$
shall be less than any assignable quantity, and consesequently less than $\frac{du}{dx}$. Then, for any value of h still smaller, we shall also have

$$\frac{du}{dx} > h\left(\frac{d^2u}{dx^2}\frac{1}{1.2} + \frac{d^3u}{dx^3}\frac{h}{1.2.3} +, \&c.\right);$$

or, if we multiply both sides of the inequality by h, we shall have

$$\frac{du}{dx}h > \frac{d^2u}{dx^2}\frac{h^2}{1.2} + \frac{d^3u}{dx^3}\frac{h^3}{1.2.3} +, \&c.:$$

that is, *when a series is expressed in the ascending powers of a variable, so small a value may be assigned to that variable as shall render the first term of the series greater than the sum of all the other terms, and this inequality will increase for all values of the variable which are still less. Under such a supposition the sign of the series will depend on that of its first term.*

45. *Remark.* The theorem of Taylor has been demonstrated under the supposition that the *form* of the function

$$u' = f(x + y),$$

is independent of the particular values which may be attributed to either of the variables x or y. Hence, when we make $y = 0$, and obtain

$$u = f(x),$$

this function of x ought to preserve the same form as $f(x + y)$; else there would be values of x in one of the functions,

$$u' = f(x+y), \qquad u = f(x),$$

which would not be found in the other, and consequently some of the values of x would be made to disappear when a particular value is assigned to y, which is entirely contrary to the supposition.

If the function be of the form

$$u' = b + \sqrt{a - x + y},$$

we shall have

$$u = b + \sqrt{a - x}.$$

If we now make $x=a$, we shall have

$$u' = b + \sqrt{y}, \quad \text{and} \quad u = b,$$

in which we see, that u' and u are expressed under different forms; and hence, the particular value of $y=0$ changes the form of the function, which is contrary to the hypothesis of Taylor's theorem. When, therefore, the function

$$u' = f(x+y),$$

shall change its form by attributing particular values to x or y, the development cannot be made by Taylor's theorem, for such particular values.

46. The particular supposition which changes the form of the function will, in general, render the differential coefficients in the development equal to infinity.

If we have

$$u' = c + \sqrt{f+x-y},$$

then,

$$u = c + \sqrt{f+x},$$

$$\frac{du}{dx} = \frac{1}{2(f+x)^{\frac{1}{2}}}$$

$$\frac{d^2u}{dx^2} = -\frac{1}{2\times 2(f+x)^{\frac{3}{2}}}$$

$$\frac{d^3u}{dx^3} = \frac{1\ .\ 3}{2\times 2\times 2(f+x)^{\frac{5}{2}}}$$

&c. &c.

in which all the coefficients will become equal to infinity when we make $x=-f$.

47. If we have a function of the form

$$u' = b + \sqrt[n]{a - x + y},$$

in which n is a whole number, all the differential coefficients of u, for $x = a$ will become infinite. For, we have

$$u = b + \sqrt[n]{a - x} = b + (a - x)^{\frac{1}{n}}$$

hence,

$$\frac{du}{dx} = -\frac{1}{n}\frac{1}{(a - x)^{\frac{n-1}{n}}},$$

$$\frac{d^2u}{dx^2} = \frac{(1 - n)}{n^2}\frac{1}{(a - x)^{\frac{2n-1}{n}}},$$

&c. &c.

all of which become infinite when we make $x = a$.

Maclaurin's Theorem.

48. MACLAURIN'S THEOREM explains the method of developing into a series any function of a single variable

Let us suppose the function to be of the form

$$u = f(x).$$

It is plain that the value of $f(x)$ must be expressed in terms of x, and of the constants which enter into $f(x)$.

Let us therefore assume

$$u = A + Bx^a + Cx^b + Dx^c +, \text{ \&c.},$$

in which the terms are arranged according to the ascending powers of x, and in which A, B, C, D, &c., are independent of x, and dependent on the constants which enter into $f(x)$.

It is now required to find such values for the exponents a, b, c, &c., and the coefficients A, B, C, D, &c., as shall render the development true for all possible values which may be attributed to x.

If we make $x=0$, u takes that value which the $f(x)$ assumes under this supposition, and if we designate that value by U we shall have

$$U=A.$$

The first differential coefficient is

$$\frac{du}{dx}=aBx^{a-1}+bCx^{b-1}+cDx^{c-1}+\&c,$$

and since this does not necessarily become 0 when we make $x=0$, it follows that there must be one term in the second member of the form x^0: hence,

$$a-1=0, \qquad \text{or} \qquad a=1;$$

and making $x=0$, we have

$$\frac{du}{dx}=B=U'$$

The second differential coefficient is

$$\frac{d^2u}{dx^2}=b(b-1)Cx^{b-2}+c(c-1)Dx^{c-2}+\&c.;$$

but since the second differential coefficient does not necessarily become 0, when $x=0$, we have

$$b-2=0, \qquad \text{or} \qquad b=2:$$

hence, by making $x=0$, we have

$$\frac{d^2u}{dx^2}=2C, \qquad \text{or} \qquad C=\frac{d^2u}{dx^2}\frac{1}{2}.=\frac{U''}{1.2}$$

We may prove in a similar manner that

$$c=3 \quad \text{and} \quad D=\frac{d^3u}{dx^3}\frac{1}{1.2.3},=\frac{U'''}{1.2.3}$$

Having designated by U what the function becomes when we make $x=0$, and by U', U'', U''', &c., what the successive differential coefficients become under the same supposition, we shall have

$$f(x)=U+U'x+U''\frac{x^2}{1.2}+U'''\frac{x^3}{1.2.3}+\&c.$$

49. The theorem of Maclaurin may be deduced immediately from that of Taylor.

In the development

$$u'=u+\frac{du}{dx}y+\frac{d^2u}{dx^2}\frac{y^2}{1.2}+\frac{d^3u}{dx^3}\frac{y^3}{1.2.3}+\&c.,$$

the coefficients $u, \quad \frac{du}{dx}, \quad \frac{d^2u}{dx^2}, \quad$ &c.,

are functions of x, and also dependent on the constants which enter into $f(x+y)$.

If we make $x=0$, the $f(x+y)$ becomes $f(y)$, and each of the differential coefficients being thus made independent of x, will depend only on the constants which enter into $f(x+y)$, and which also enter into $f(y)$ Hence, if we designate by

$$U, \ U', \ U'', \ U''', \ U'''', \ \&c.,$$

the values which the coefficients assume under this hypothesis, we shall have

$$f(y) = U + U'y + U''\frac{y^2}{1.2} + U'''\frac{y^3}{1.2.3} + U''''\frac{y^4}{1.2.3.4} + \&c.$$

50. If we take a function of the form

$$u = (a + x)^n,$$

we shall have

$$\frac{du}{dx} = n(a + x)^{n-1},$$

$$\frac{d^2u}{dx^2} = n(n-1)(a + x)^{n-2},$$

$$\&c. = \&c.$$

which become, when we make $x = 0$,

$$U = a^n, \quad U' = na^{n-1}, \quad U'' = n(n-1)a^{n-2}, \quad \&c.;$$

hence,

$$(a + x)^n = a^n + na^{n-1}x + \frac{n(n-1)}{1.2}a^{n-2}x^2 + \&c.$$

51. *Remark* 1. The theorem of Maclaurin has been demonstrated under the supposition that the $f(x)$ reduces to a finite quantity when we make $x = 0$. The case, therefore, is excluded in which $x = 0$ renders the function infinite. Thus, if we have

$$u = \cot x, \quad u = \operatorname{cosec} x, \quad \text{or} \quad u = \log x,$$

and make $x = 0$, we find $u = \infty$; hence, neither of these functions can be developed by the theorem of Maclaurin.

Remark 2. We have already seen (Art. 45.), that the theorem of Taylor does not apply to those cases in which the form of the function is changed by attributing a *particular* value to one of the variables: the theorem therefore *fails* for *particular* values, but is true for all others, and hence, the *general* development never fails.

In the theorem of Maclaurin the failure arises from the *form* of the function: hence, it is the *general* development which fails, and with it, all the particular cases.

EXAMPLES.

1. Develop into a series the function

$$u = \sqrt{a^2 + x^2} = a\left(1 + \frac{x^2}{a^2}\right)^{\frac{1}{2}}.$$

2. Develop into a series the function

$$u = \sqrt[3]{(a^2 - x^2)^2} = a^{\frac{4}{3}}\left(1 - \frac{x^2}{a^2}\right)^{\frac{2}{3}}.$$

3. Develop into a series the function

$$u = \frac{1}{a + x} = a^{-1}\left(1 + \frac{x}{a}\right)^{-1}.$$

4. Develop into a series the function

$$u = \frac{1}{\sqrt[4]{a^4 + x^4}} = a^{-1}\left(1 + \frac{x^4}{a^4}\right)^{-\frac{1}{4}}$$

CHAPTER III.

Of Transcendental Functions.

52. If we have an equation of the form

$$u = a^x,$$

in which a is constant, it is plain that u will be a function of x; and if a be made the base of a system of logarithms, x will be the logarithm of the number u (Alg. Art, 257). When the variable and function are thus related to each other, u is said to be an *exponential or logarithmic function of* x. (Art. 9).

53. The functions expressed by the equations

$$u = \sin x, \quad u = \cos x, \quad u = \text{tang } x, \quad u = \cot x, \quad \&c.,$$

are called *circular functions.*

The logarithmic and circular functions are generally called *transcendental functions,* because the relation between the function and variable is not determined by the ordinary operations of Algebra.

Differentiation of Logarithmic Functions.

54. Let us resume the function

$$u = a^x.$$

If we give to x an increment h, we have

$$u' = a^{x+h},$$

and $$u' - u = a^{x+h} - a^x = a^x(a^h - 1).$$

In order to develop a^h, let us make $a = 1 + b$, we shall then have

$$a^h = (1+b)^h = 1 + \frac{h}{1}b + \frac{h(h-1)}{1.2}b^2 + \frac{h(h-1)(h-2)}{1.2.3}b^3 + \text{\&c};$$

hence,

$$a^h - 1 = \frac{h}{1}b + \frac{h(h-1)}{1.2}b^2 + \frac{h(h-1)(h-2)}{1.2.3}b^3 + \text{\&c.},$$

$$= h\left(\frac{b}{1} + \frac{(h-1)}{1}\frac{b^2}{2} + \frac{(h-1)(h-2)}{1.2}\frac{b^3}{3} + \text{\&c.}\right);$$

from which we see, that the coefficients of the first power of h will be

$$\left(\frac{b}{1} - \frac{b^2}{2} + \frac{b^3}{3} - \text{\&c.}\right);$$

replacing b by its value $a - 1$, and passing to the limit, we obtain

$$\frac{du}{dx} = \frac{da^x}{dx} = a^x\left(\frac{a-1}{1} - \frac{(a-1)^2}{2} + \frac{(a-1)^3}{3} - \text{\&c.}\right)$$

or if we make

$$k = \frac{a-1}{1} - \frac{(a-1)^2}{2} + \frac{(a-1)^3}{3} - \text{\&c.},$$

$$\frac{da^x}{dx} = ka^x, \quad \text{or} \quad da^x = ka^x dx;$$

in which k is dependent on a.

The successive differential coefficients are readily found. For we have

$$\frac{da^x}{dx} = a^x k,$$

$$d\left(\frac{da^x}{dx}\right) = da^x k = a^x k^2 dx;$$

hence,

$$\frac{d^2a^x}{dx^2} = a^x k^2,$$

$$\frac{d^3a^x}{dx^3} = a^x k^3,$$

$$\&c. \qquad \&c$$

$$\frac{d^na^x}{dx^n} = a^x k^n.$$

55. It is now proposed to find the relation which exists between a and k. For this purpose, let us employ the formula of Maclaurin,

$$u = f(x) = U + U'\frac{x}{1} + U''\frac{x^2}{1.2} + U'''\frac{x^3}{1.2.3} + \ \&c.$$

If in the function

$$u = a^x,$$

and the successive differential coefficients before found, we make $x = 0$, we have

$$U = 1, \qquad U' = k, \qquad U'' = k^2, \qquad U''' = k^3, \quad \&c.;$$

hence,

$$a^x = 1 + \frac{kx}{1} + \frac{k^2x^2}{1.2} + \frac{k^3x^3}{1.2.3} + \ \&c.$$

If we now make $x = \frac{1}{k}$, we shall have

$$a^{\frac{1}{k}} = 1 + \frac{1}{1} + \frac{1}{1.2} + \frac{1}{1.2.3} + \ \&c.;$$

designating by e the second member of the equation, and employing twelve terms of the series, we shall find

$$e = 2.7182818;$$

hence, $$a^{\frac{1}{k}} = e, \quad \text{therefore} \quad a = e^k.$$

But, 2.7182818 is the base of the Naperian system of logarithms (Alg. Art. 272); hence, *the constant quantity* k *is the Naperian logarithm of* a.

By resuming the result obtained in Art. 54,

$$da^x = a^x k\, dx,$$

we see that *the differential of a quantity obtained by raising a constant to a power denoted by a variable exponent, is equal to the quantity itself into the Naperian logarithm of the constant, into the differential of the exponent.*

56. If now we take the logarithms, in any system, of both members of the equation

$$e^k = a,$$

we shall have

$$kle = la, \quad \text{or} \quad k = \frac{la}{le},$$

whence,

$$da^x = ka^x dx = \frac{la}{le} a^x dx;$$

or by recollecting that

$$u = a^x,$$

we have

$$\frac{du}{dx} = \frac{la}{le} a^x;$$

or, if we regard x as the function, and u as the variable, we have (Art. 34),

$$\frac{dx}{du} = \frac{le}{la}\frac{1}{a^x}.$$

Let us now suppose a to be the base of a system of logarithms. We shall then have $x =$ the logarithm of u, $la = 1$, and $le =$ the modulus of the system (Alg. Art. 272); and the equation will become

$$d(lu) = le\frac{du}{u},$$

that is, *the differential of the logarithm of a quantity is equal to the modulus of the system into the differential of the quantity divided by the quantity itself.*

57. If we suppose $a = e$ the base of the Naperian system, and employ the usual characteristic l' to designate the Naperian logarithm, we shall have

$$d(l'u) = \frac{du}{u};$$

that is, *the differential of the Naperian logarithm of a quantity is equal to the differential of the quantity divided by the quantity itself.*

The last property might have been deduced from the preceding article by observing that the modulus of the Naperian system is equal to unity.

58. The theorem of Maclaurin affords an easy method of finding a logarithmic series from which a table of logarithms may be computed. If we have a function of the form,

$$u = f(x) = lx,$$

we have already seen that the development cannot be made, since $f(x)$ becomes infinite when $x=0$ (Art. 51.)

But if we make

$$u=f(x)=l(1+x),$$

the function will not become infinite when $x=0$; and hence the development may be made.

The theorem of Maclaurin gives

$$u=f(x)=U+U'\frac{x}{1}+U''\frac{x^2}{1.2}+U'''\frac{x^3}{1.2.3}+\&c.$$

If we designate the modulus of the system of the logarithms by A, we shall have

$$\frac{du}{dx}=A\frac{1}{1+x}=A(1+x)^{-1},$$

$$\frac{d^2u}{dx^2}=-A\frac{1}{(1+x)^2}=-A(1+x)^{-2},$$

$$\frac{d^3u}{dx^3}=2A\frac{1}{(1+x)^3}=2A(1+x)^{-3}.$$

If we now make $x=0$, we have

$$U=0,\quad U'=A,\quad U''=-A,\quad U'''=2A,\ \&c.;$$

hence,

$$l(1+x)=A\left(x-\frac{x^2}{2}+\frac{x^3}{3}-\frac{x^4}{4}+\frac{x^5}{5}-\&c.\right)$$

This series is not sufficiently converging, except in the case when x is a very small fraction. To render the series more converging, substitute $-x$ for x: we then have

$$l(1-x)=A\left(-x-\frac{x^2}{2}-\frac{x^3}{3}-\frac{x^4}{4}-\frac{x^5}{5}-\&c.\right)$$

and by subtracting the last series from the first, we obtain

$$l(1+x)-l(1-x)=l\left(\frac{1+x}{1-x}\right)=2A\left(\frac{x}{1}+\frac{x^3}{3}+\frac{x^5}{5}+\text{ \&c.}\right)$$

If we make

$$\frac{1+x}{1-x}=1+\frac{z}{n}, \quad \text{we have} \quad x=\frac{z}{2n+z},$$

and by observing that

$$l\left(1+\frac{z}{n}\right)=l\left(\frac{n+z}{n}\right)=l(n+z)-ln,$$

we have

$$l(n+z)-ln=2A\left[\frac{z}{2n+z}+\frac{1}{3}\left(\frac{z}{2n+z}\right)^3+\frac{1}{5}\left(\frac{z}{2n+z}\right)^5+\text{ \&c.,}\right]$$

from which we can find the logarithm of $n+z$ when the logarithm of n is known. This series is similar to that found in Algebra, Art. 270.

If we make $n=1$, and $z=1$, we have $l1=0$, and

$$l2=2A\left(\frac{1}{3}+\frac{1}{3.3^3}+\frac{1}{5.3^5}+\text{ \&c.}\right)$$

If we make the modulus $A=1$, the logarithm will be taken in the Naperian system, and we shall have

$$l'2=0.693147180,$$

$$2l'2=l'4=1.386294360;$$

and by making $z=4$, and $n=1$, we have

$$l'5=1.609437913,$$

and $\quad l'2+l'5=l'10=2.302585093.$

If we now suppose the first logarithms to have been taken in the common system, of which the base is 10, we shall have, by recollecting, that the logarithms of the same number taken in two different systems are to each other as their moduli (Alg. Art. 267),

$$l10 \ : \ l'10 \ :: \ A \ : \ 1,$$

or,

$$1 \ : \ 2.302585093 \ :: \ A \ : \ 1;$$

whence,

$$A = \frac{1}{2.30258509} = 0.434284482.$$

Remark. To avoid the inconvenience of writing the modulus at each differentiation (Art. 56), the Naperian logarithms are generally used in the calculus, and when we wish to pass to the common system, we have merely to multiply by the modulus of the common system. We may then omit the accent, and designate the Naperian logarithm by l.

59. Let us now apply these principles in differentiating logarithmic functions.

1. Let us take the function $\quad u = l\left(\dfrac{x}{\sqrt{a^2+x^2}}\right).$

Make

$$z = \frac{x}{\sqrt{a^2+x^2}},$$

and we shall have

$$du = \frac{dz}{z},$$

but

$$dz = \frac{dx\,\sqrt{a^2+x^2} - \dfrac{x^2dx}{\sqrt{a^2+x^2}}}{a^2+x^2} = \frac{a^2dx}{\left(a^2+x^2\right)^{\frac{3}{2}}}$$

whence, $$du = \frac{a^2 dx}{x(a^2+x^2)}.$$

2. Take the function $u = l\left[\frac{\sqrt{1+x}+\sqrt{1-x}}{\sqrt{1+x}-\sqrt{1-x}}\right]$;

and make $\sqrt{1+x}+\sqrt{1-x}=y$, $\sqrt{1+x}-\sqrt{1-x}=z$,
which gives

$$u = l\left(\frac{y}{z}\right) = ly - lz, \quad \text{and} \quad du = \frac{dy}{y} - \frac{dz}{z}.$$

But we have

$$dy = \frac{dx}{2\sqrt{1+x}} - \frac{dx}{2\sqrt{1-x}} = \frac{-dx}{2\sqrt{1-x^2}}\left(\sqrt{1+x}-\sqrt{1-x}\right),$$

$$= -\frac{zdx}{2\sqrt{1-x^2}},$$

$$dz = \frac{dx}{2\sqrt{1+x}} + \frac{dx}{2\sqrt{1-x}} = \frac{dx}{2\sqrt{1-x^2}}\left(\sqrt{1+x}+\sqrt{1-x}\right),$$

$$= \frac{ydx}{2\sqrt{1-x^2}}.$$

Whence,

$$\frac{dy}{y} - \frac{dz}{z} = -\frac{zdx}{2y\sqrt{1-x^2}} - \frac{ydx}{2z\sqrt{1-x^2}},$$

$$= \frac{-(y^2+z^2)dx}{2yz\sqrt{1-x^2}};$$

and observing that $y^2+z^2=4$ and $yz=2x$,

we have $$du = -\frac{dx}{x\sqrt{1-x^2}}.$$

3. $u = l\left(x + \sqrt{1+x^2}\right), \qquad du = \frac{dx}{\sqrt{1+x^2}}.$

4. $u = \frac{1}{\sqrt{-1}}\, l\left(x\sqrt{-1} + \sqrt{1-x^2}\right), \quad du = \frac{dx}{\sqrt{1-x^2}}.$

5. $u = l\left[\frac{\sqrt{1+x^2}+x}{\sqrt{1+x^2}-x}\right]^{\frac{1}{2}}, \qquad du = \frac{dx}{\sqrt{1+x^2}}.$

6. $u = l\left[\frac{\sqrt{a+x}+\sqrt{a-x}}{\sqrt{a+x}-\sqrt{a-x}}\right], \qquad du = -\frac{adx}{x\sqrt{a^2-x^2}}.$

60. Let us suppose that we have a function of the form

$$u = (lx)^n.$$

Make $lx = z$, and we have

$$u = z^n, \qquad du = nz^{n-1}dz,$$

and substituting for z and dz their values,

$$d(lx)^n = \frac{n(lx)^{n-1}}{x}dx.$$

61. Let us suppose that we have

$$u = l(lx).$$

Make $lx = z$, and we shall have,

$$u = lz, \qquad du = \frac{dz}{z}, \qquad dz = \frac{dx}{x};$$

hence, - - $du = \frac{dx}{xlx}.$

62. The rules for the differentiation of logarithmic functions are advantageously applied in the differentiation of complicated exponential functions.

1. Let us suppose that we have a function of the form

$$u = z^y,$$

in which z and y are both variables.

If we take the logarithms of both members, we have

$$lu = ylz\,;$$

hence,
$$\frac{du}{u} = dy\,lz + y\frac{dz}{z}\,;$$

or,
$$du = ulzdy + uy\frac{dz}{z},$$

or by substituting for u its value

$$du = dz^y = z^y lzdy + yz^{y-1}dz.$$

Hence, *the differential of a function which is equal to a variable root raised to a power denoted by a variable exponent, is equal to the sum of the differentials which arise, by differentiating, first under the supposition that the root remains constant, and then under the supposition that the exponent remains constant* (Arts. 55, and 32).

2. Let the function be of the form

$$u = a^{b^x}.$$

Make, $b^x = y$, and we shall then have (Art. 55),

$$u = a^y, \qquad du = a^y lady\,; \quad \text{but} \quad dy = b^x lbdx,$$

hence,
$$du = a^{b^x} b^x lalbdx.$$

3. Let us take as a last example

$$u = z^{t^s},$$

in which z, t, and s, are variables.

Make, $t^s = y$, we shall then have

$$u = z^y, \qquad du = z^y lz dy + y z^{y-1} dz.$$

But $\quad dy = t^s lt\, ds + s t^{s-1} dt;$

hence, $\quad du = z^{t^s} lz (t^s lt\, ds + s t^{s-1} dt) + t^s z^{t^s - 1} dz,$

$$du = z^{t^s} t^s \left(lt\, lz\, ds + \frac{slzdt}{t} + \frac{dz}{z} \right).$$

Differentiation of Circular Functions.

63. Let us first find the differential of the sine of an arc. For this purpose we will assume the formulas (Trig. Art XIX),

$$\sin(a+b) = \frac{\sin a \cos b + \sin b \cos a}{R},$$

$$\sin(a-b) = \frac{\sin a \cos b - \sin b \cos a}{R}.$$

If we subtract the second equation from the first,

$$\sin(a+b) - \sin(a-b) = \frac{2 \sin b \cos a}{R}.$$

and if we make $a + b = x + h$, and $a - b = x$, we shall have

$$\sin(x+h) - \sin x = \frac{2 \sin \frac{1}{2} h \cos \left(x + \frac{1}{2} h \right)}{R},$$

and dividing both members by h,

$$\frac{\sin(x+h)-\sin x}{h}=\frac{2\sin\frac{1}{2}h\cos\left(x+\frac{1}{2}h\right)}{hR},$$

$$=\frac{\sin\frac{1}{2}h}{\frac{1}{2}h}\cdot\frac{\cos\left(x+\frac{1}{2}h\right)}{R}.$$

If we now pass to the limit, the second factor of the second member of the equation will become $\frac{\cos x}{R}$.

In relation to the first factor $\frac{\sin\frac{1}{2}h}{\frac{1}{2}h}$ its limit will be unity.

For, $\quad \text{tang}\, a=\frac{R\sin a}{\cos a}$, whence $\frac{\sin a}{\text{tang}\, a}=\frac{\cos a}{R}$;

Now, since an arc is greater than its sine and less than its tangent*

$$\frac{\sin a}{a}<1, \quad \text{and} \quad \frac{\sin a}{a}>\frac{\sin a}{\text{tang}\, a}:$$

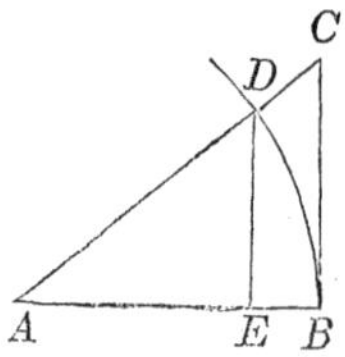

* The arc DB is greater than a straight line drawn from D to B, and consequently greater than the sine DE drawn perpendicular to AB.

The area of the sector ABD is equal to $\frac{1}{2}AB\times BD$, and the area of the triangle ABC is equal to $\frac{1}{2}AB\times BC$. But the sector is less than the triangle being contained within it: hence,

C
D
A
E
B

$$\frac{1}{2}AB\times BD<\frac{1}{2}AB\times BC,$$

consequently, $\qquad BD < BC.$

hence, the ratio of the sine divided by the arc is nearer unity than that of the sine divided by the tangent. But when we pass to the limit, by making the arc equal to 0, the sine divided by the tangent being equal to the cosine divided by the radius, is equal to unity: hence *the limit of the ratio of the sine and arc, is unity.*

When therefore we pass to the limit by making $h=0$, we find

$$\frac{d\sin x}{dx}=\frac{\cos x}{R}:$$

hence,
$$d\sin x=\frac{\cos x\,dx}{R}.$$

64. Having found the differential of the sine, the differentials of the other functions of the arc are readily deduced from it.

$$\cos x=\sin(90^\circ-x), \qquad d\cos x=d\sin(90^\circ-x),$$

and by the last article,

$$d\sin(90^\circ-x)=\frac{1}{R}\cos(90^\circ-x)\,d(90^\circ-x),$$

$$=-\frac{1}{R}\cos(90^\circ-x)dx:$$

hence,
$$d\cos x=-\frac{\sin x\,dx}{R};$$

the differential of the cosine in terms of the arc bei , negative, as it should be, since the cosine and arc are lecreasing functions of each other (Art. 31.)

65. Since the versed sine of an arc is equal to radius minus the cosine, we have

$$d\,\text{ver-sin}\,x = d(R - \cos x) = \frac{\sin x dx}{R}.$$

66. Since $\tan x = \dfrac{R \sin x}{\cos x}$, we have (Art. 30),

$$d\,\text{tang}\,x = \frac{R\cos x d\sin x - R\sin x d\cos x}{\cos^2 x},$$

$$= \frac{(\cos^2 x + \sin^2 x)dx}{\cos^2 x};$$

but $$\cos^2 x + \sin^2 x = R^2:$$

hence, $$d\,\text{tang}\,x = \frac{R^2 dx}{\cos^2 x}.$$

67. Since $\cot x = \dfrac{R^2}{\text{tang}\,x}$, we have

$$d\cot x = -\frac{R^2 d\,\text{tang}\,x}{\text{tang}^2 x} = -\frac{R^4 dx}{\text{tang}^2 x \cos^2 x};$$

but, $$\text{tang}^2 x = \frac{R^2 \sin^2 x}{\cos^2 x};$$

hence, $$d\cot x = -\frac{R^2 dx}{\sin^2 x};$$

which is negative, as it should be, since the cotangent is a decreasing function of the arc.

68. Since $\sec x = \frac{R^2}{\cos x}$, we have

$$d \sec x = -\frac{R^2 d \cos x}{\cos^2 x} = \frac{R \sin x \, dx}{\cos^2 x}:$$

but, $\frac{R \sin x}{\cos x} = \text{tang} x$, and $\frac{R^2}{\cos x} = \sec x$;

hence, $$d \sec x = \frac{\sec x \, \text{tang} x \, dx}{R^2}.$$

69. Since $\text{cosec} x = \frac{R^2}{\sin x}$, we have

$$d \text{ cosec } x = -\frac{R^2 d \sin x}{\sin^2 x} = -\frac{R \cos x \, dx}{\sin^2 x};$$

hence, $$d \text{ cosec} x = -\frac{\text{cosec} x \cot x \, dx}{R^2}.$$

70. If we make $R = 1$, Arts. 63, 64, 65, 66, 67, will give,

$$d \sin x = \cos x \, dx \qquad (1),$$

$$d \cos x = - \sin x \, dx \qquad (2),$$

$$d \text{ ver} \sin x = \sin x \, dx \qquad (3),$$

$$d \text{ tang} x = \frac{dx}{\cos^2 x} \qquad (4),$$

$$d \cot x = -\frac{dx}{\sin^2 x} \qquad (5).$$

The differential values of the secant and cosecant are omitted, being of little practical use.

71. In treating the circular functions, it is found to be most convenient to regard the arc as the function, and the

sine, cosine, versed-sine, tangent, or cotangent, as the variable. If we designate the variable by u, we shall have in (Art. 63) $\sin x = u$, and

$$dx = \frac{Rdu}{\cos x} = \frac{Rdu}{\sqrt{R^2 - u^2}}.$$

If we make $\cos x = u$, we have (Art. 64),

$$dx = -\frac{Rdu}{\sin x} = -\frac{Rdu}{\sqrt{R^2 - u^2}}.$$

If we make $\text{ver-sin}\, x = u$, we have (Art. 65),

$$dx = \frac{Rdu}{\sin x}.$$

But, $\sin x = \sqrt{R^2 - \cos^2 x}$, and $\cos x = R - u$,

therefore, $\cos^2 x = R^2 - 2Ru + u^2$,

hence, $\sin x = \sqrt{2Ru - u^2}$,

and consequently, $dx = \dfrac{Rdu}{\sqrt{2Ru - u^2}}$;

If we make $\tan x = u$, we have (Art. 66)

$$dx = \frac{\cos^2 x\, du}{R^2};$$

but $\dfrac{\cos x}{R} = \dfrac{R}{\sec x}$, hence $\dfrac{\cos^2 x}{R^2} = \dfrac{R^2}{\sec^2 x} = \dfrac{R^2}{R^2 + \tan^2 x}$,

hence, $dx = \dfrac{R^2 du}{R^2 + u^2}$.

Now, if we make $R=1$, the four last formulas become

$$dx=\frac{du}{\sqrt{1-u^2}}, \qquad dx=-\frac{du}{\sqrt{1-u^2}},$$

$$dx=\frac{du}{\sqrt{2u-u^2}}, \qquad dx=\frac{du}{1+u^2};$$

and these formulas being of frequent use, should be carefully committed to memory.

72. The following notation has recently been introduced into the differential calculus, and it enables us to designate an arc by means of its functions.

$\sin^{-1}u=$ the arc of which u is the sine,

$\cos^{-1}u=$ the arc of which u is the cosine,

$\text{tang}^{-1}u=$ the arc of which u is the tangent,

&c. &c. &c.

If, for example, we have

$$x=\sin^{-1}u, \quad \text{then,} \quad dx=\frac{du}{\sqrt{1-u^2}}.$$

73. We shall now add a few examples.

1. Let us take a function of the form

$$x=\cos x^{\sin x}.$$

Make $\cos x=z$, and $\sin x=y$;

then, $u=z^y$, and (Art. 62);

$$du=z^y\, lz\, dy+yz^{y-1}dz:$$

also, $dz = -\sin x\,dx$, and $dy = \cos x\,dx$.

hence, $$du = z^y\left(lz\,dy + \frac{y}{z}\,dz\right),$$

$$= \cos x^{\sin x}\left(l\cos x\cos x - \frac{\sin^2 x}{\cos x}\right)dx.$$

2. Differentiate the function

$$x = \sin^{-1} mu, \qquad dx = \frac{m\,du}{\sqrt{1-m^2u^2}}$$

3. Differentiate the function

$$x = \cos^{-1}\left(u\sqrt{1-u^2}\right)$$

$$dx = \frac{(-1+2u^2)\,du}{\sqrt{(1-u^2+u^4)(1-u^2)}}.$$

4. Differentiate the function

$$x = \text{tang}^{-1}\frac{u}{2}, \qquad dx = \frac{2\,du}{4+u^2}.$$

5. Differentiate the function

$$x = \sin^{-1}\left(2u\sqrt{1-u^2}\right), \qquad dx = \frac{2\,du}{\sqrt{1-u^2}}.$$

6. Differentiate the function

$$u = \text{tang}^{-1}\frac{x}{y}, \qquad du = \frac{y\,dx - x\,dy}{y^2+x^2}.$$

74. We are enabled by means of Maclaurin's theorem and the differentials of the circular functions, to find the

value of the principal functions of an arc in terms of the arc itself.

Let $u = f(x) = \sin x$: then,

$$\frac{du}{dx} = \cos x, \quad \frac{d^2u}{dx^2} = -\sin x, \quad \frac{d^3u}{dx^3} = -\cos x,$$

$$\frac{d^4u}{dx^4} = \sin x, \quad \frac{d^5u}{dx^5} = +\cos x.$$

If we now render the differential coefficients independent of x, by making $x = 0$, we have (Art. 49),

$$U = 0, \quad U' = 1, \quad U'' = 0, \quad U''' = -1,$$

$$U'''' = 0, \quad U''''' = +1:$$

hence,
$$\sin x = \frac{x}{1} - \frac{x^3}{1.2.3} + \frac{x^5}{1.2.3.4.5} - \&c.$$

75. To develop the cosine in terms of the arc, make

$$u = f(x) = \cos x; \quad \text{then,}$$

$$\frac{du}{dx} = -\sin x, \quad \frac{d^2u}{dx^2} = -\cos x, \quad \frac{d^3u}{dx^3} = \sin x,$$

$$\frac{d^4u}{dx^4} = \cos x, \quad \frac{d^5u}{dx^5} = -\sin x,$$

and rendering the coefficients independent of x, we have

$$U = 1, \quad U' = 0, \quad U'' = -1, \quad U''' = 0,$$

$$U'''' = 1, \quad U''''' = 0:$$

hence,
$$\cos x = 1 - \frac{x^2}{1.2} + \frac{x^4}{1.2.3.4} - \&c.$$

The last two formulas are very convenient in calculating the trigonometrical tables, and when the arc is small the series will converge rapidly. Having found the sine and cosine, the other functions of the arc may readily be calculated from them.

76. In the two last series we have found the values of the functions, sine and cosine, in terms of the arc. We may, if we please, find the value of the arc in terms of any of its functions.

77. The differential coefficient of the arc in terms of its sine, is (Art. 71),

$$\frac{dx}{du} = \frac{1}{\sqrt{1-u^2}} = (1-u^2)^{-\frac{1}{2}};$$

developing by the binomial formula, we find

$$\frac{dx}{du} = 1 + \frac{1}{2}u^2 + \frac{1.3}{2.4}u^4 + \frac{1.3.5}{2.4.6}u^6 + \&c.$$

In passing from the function to the differential coefficient, the exponent of the variable in each term which contains it, is diminished by unity; and hence, the series which expresses the value of x in terms of u, will contain the uneven powers of u, or will be of the form

$$x = Au + Bu^3 + Cu^5 + Du^7 + \&c.;$$

and the differential coefficient is

$$\frac{dx}{du} = A + 3Bu^2 + 5Cu^4 + 7Du^6 + \&c.$$

But since the differential coefficients are equal to each other, we find, by comparing the series,

$$A=1, \quad B=\frac{1}{2.3}, \qquad C=\frac{1.3}{2.4.5}, \qquad D=\frac{1.3.5}{2.4.6.7};$$

hence,

$$x=\sin^{-1}u=\frac{u}{1}+\frac{1}{2}\frac{u^3}{3}+\frac{1.3u^5}{2.4.5}+\frac{1.3.5}{2.4.6.7}u^7+\ \&c.$$

If we take the arc of 30°, of which the sine is $\frac{1}{2}$ (Trig. Art. XV), we shall have

$$\text{arc } 30° = \frac{1}{2}+\frac{1}{2.3.2^3}+\frac{1.3}{2.4.5.2^5}+\frac{1.3.5}{2.4.6.7.2^7}+\ \&c.,$$

and by multiplying both members of the equation by 6, we obtain the length of the semi-circumference to the radius unity.

78. To express the arc in terms of its tangent, we have (Art. 71),

$$\frac{dx}{du}=\frac{1}{1+u^2}=(1+u^2)^{-1},$$

which gives

$$\frac{dx}{du}=1-u^2+u^4-u^6+\ \&c.;$$

hence the function x must be of the form

$$x=Au+Bu^3+Cu^5+Du^7,$$

and consequently

$$\frac{dx}{du}=A+3Bu^2+5Cu^4+7Du^6;$$

and by comparing the series, and substituting for A, B, C, &c., their values, we find

$$x = \text{tang}^{-1} u = \frac{u}{1} - \frac{u^3}{3} + \frac{u^5}{5} - \frac{u^7}{7} + \&c.$$

If we make $x = 45^\circ$, u will be equal to 1; hence,

$$\text{arc } 45^\circ = 1 - \frac{1}{3} + \frac{1}{5} - \frac{1}{7} + \&c.$$

But this series is not sufficiently convergent to be used for computing the value of the arc. To find the value of the arc in a more converging series, we employ the following property of two arcs, viz.:

Four times the arc whose tangent is $\frac{1}{5}$, *exceeds the arc of* 45° *by the arc whose tangent is* $\frac{1}{239}$*.

---

* Let a represent the arc whose tangent is $\frac{1}{5}$. Then (Trig. Art. XXVI),

$$\text{tang } 2a = \frac{2 \text{ tang } a}{1 - \text{tang}^2 a} = \frac{5}{12},$$

$$\text{tang } 4a = \frac{2 \text{ tang } 2a}{1 - \text{tang}^2 2a} = \frac{120}{119}.$$

The last number being greater than unity, shows that the arc $4a$ exceeds 45°. Making

$$4a = A, \qquad 45^\circ = B,$$

the difference, $4a - 45^\circ = A - B = b$, will have for its tangent

$$\text{tang } b = \text{tang } (A - B) = \frac{\text{tang } A - \text{tang } B}{1 + \text{tang } A \text{ tang } B} = \frac{1}{239};$$

hence, *four times the arc whose tangent is* $\frac{1}{5}$, *exceeds the arc of* 45° *by an arc whose tangent is* $\frac{1}{239}$.

But $$\tan^{-1}\frac{1}{5} = \frac{1}{5} - \frac{1}{3.5^3} + \frac{1}{5.5^5} - \frac{1}{7.5^7} + \&c.,$$

$$\tan^{-1}\frac{1}{239} = \frac{1}{239} - \frac{1}{3(239)^3} + \frac{1}{5(239)^5} - \frac{1}{7(239)^7} + \&c.;$$

hence,

$$\text{arc } 45^\circ = \left\{ \begin{array}{l} 4\left(\frac{1}{5} - \frac{1}{3.5^3} + \frac{1}{5.5^5} - \frac{1}{7.7^7} + \right) \\ -\left(\frac{1}{239} - \frac{1}{3(239)^3} + \frac{1}{5(239)^5} - \frac{1}{7(239)^7} + \right) \end{array} \right\}$$

Multiplying by 4, we find the semi-circumference

$$= 3.141592653.$$

CHAPTER IV.

Development of any Function of two Variables —Differential of a Function of any number of Variables—Implicit Functions—Differential Equations of Curves—Of Vanishing Fractions.

79. We have explained in Taylor's theorem the method of developing into a series any function of the sum or difference of two variables.

We now propose to give a general theorem of which that is a particular case, viz :

To develop into a series any function of two or more variables, when each shall have received an increment, and to find the differential of the function.

80. Before making the development it will be necessary to explain a notation which has not yet been used.

If we have a function of two variables, as

$$u = f(x, y),$$

we may suppose one to remain constant, and differentiate the function with respect to the other.

Thus, if we suppose y to remain constant, and x to vary, the differential coefficient will be

$$\frac{du}{dx} = f'(x, y); \quad (1),$$

and if we suppose x to remain constant and y to vary, the differential coefficient will be

$$\frac{du}{dy} = f''(x,y). \qquad (2).$$

The differential coefficients which are obtained under these suppositions, are called *partial differential coefficients.* The first is the partial differential coefficient with respect to x, and the second with respect to y.

81. If we multiply both members of equation (1) by dx, and both members of equation (2) by dy, we obtain

$$\frac{du}{dx}dx = f'(x,y)dx, \quad \text{and} \quad \frac{du}{dy}dy = f''(x,y)dy.$$

The expressions,

$$\frac{du}{dx}dx, \qquad \frac{du}{dy}dy,$$

are called *partial differentials;* the first a partial differential with respect to x, and the second a partial differential with respect to y: hence,

A partial differential coefficient is the differential coefficient of a function of two or more variables, under the supposition that only one of them has changed its value: and,

A partial differential is the differential of a function of two or more variables, under the supposition that only one of them has changed its value.

82. If we differentiate equation (1) under the supposition that x remains constant and y varies, we shall have

$$\frac{d\left(\frac{du}{dx}\right)}{dy} = f'''(x,y),$$

and since x and dx are constant

$$d\left(\frac{du}{dx}\right)=\frac{d(du)}{dx},$$

which we designate by

$$\frac{d^2u}{dx}:$$

hence,
$$\frac{d^2u}{dx\,dy}=f'''(x,y).$$

The first member of this equation expresses that the function u has been differentiated twice, once with respect to x, and once with respect to y.

If we differentiate again, regarding x as the variable, we obtain

$$\frac{d^3u}{dx^2dy}=f^{\text{IV}}(x,y),$$

which expresses that the function has been differentiated twice with respect to x and once with respect to y. And generally

$$\frac{d^{n+m}u}{dx^n dy^m},$$

indicates that the function u has been differentiated $n+m$ times, n times with respect to x, and m times with respect to y.

83. Resuming the function

$$u=f(x,y),$$

if we suppose y to remain constant, and give to x an arbitrary increment h, we shall have from the theorem of Taylor,

$$f(x+h,y)=u+\frac{du}{dx}\frac{h}{1}+\frac{d^2u}{dx^2}\frac{h^2}{1.2}+\frac{d^3u}{dx^3}\frac{h^3}{1.2.3}+\text{\&c.},$$

6

in which, $u, \quad \frac{du}{dx}, \quad \frac{d^2u}{dx^2}, \quad \frac{d^3u}{dx^3},$

are functions of x and y, and dependent on the constants which enter the $f(x,y)$.

If we now attribute to y an increment k, the function u, which depends on y, will become

$$u + \frac{du}{dy}k + \frac{d^2u}{dy^2}\frac{k^2}{1.2} + \frac{d^3u}{dy^3}\frac{k^3}{1.2.3} + \&c.;$$

and the function $\frac{du}{dx}$ will become

$$\frac{du}{dx} + \frac{d^2u}{dx\,dy}\frac{k}{1} + \frac{d^3u}{dx\,dy^2}\frac{k^2}{1.2} + \frac{d^4u}{dx\,dy^3}\frac{k^3}{1.2.3} + \&c.,$$

and the function $\frac{d^2u}{dx^2}$, will become

$$\frac{d^2u}{dx^2} + \frac{d^3u}{dx^2dy}\frac{k}{1} + \frac{d^4u}{dx^2dy^2}\frac{k^2}{1.2} + \frac{d^5u}{dx^2dy^3}\frac{k^3}{1.2.3} + \&c.,$$

and the function $\frac{d^3u}{dx^3}$, will become

$$\frac{d^3u}{dx^3} + \frac{d^4u}{dx^3dy}\frac{k}{1} + \frac{d^5u}{dx^3dy^2}\frac{k^2}{1.2} + \frac{d^6u}{dx^3dy^3}\frac{k^3}{1.2.3} + \&c.,$$

&c. &c. &c. &c.

Substituting these values in the development of

$$f(x+h,\ y),$$

and arranging the terms, we have

$$f(x+h, y+k)=u+\frac{du}{dy}\frac{k}{1}+\frac{d^2u}{dy^2}\frac{k^2}{1.2}+\frac{d^3u}{dy^3}\frac{k^3}{1.2.3}+ \&c.,$$

$$+\frac{du}{dx}\frac{h}{1}+\frac{d^2u}{dx\,dy}\frac{hk}{1.1}+\frac{d^3u}{dx\,dy^2}\frac{hk^2}{1.2}+ \quad \&c.,$$

$$+\frac{d^2u}{dx^2}\frac{h^2}{1.2}+\frac{d^3u}{dx^2dy}\frac{h^2k}{1.2}+ \quad \&c.,$$

$$+\frac{d^3u}{dx^3}\frac{h^3}{1.2.3}+\&c.;$$

which is the general development of a function of two variables, when each has received an increment, in terms of the increments and differential coefficients.

84. If we transpose $u=f(x, y)$ into the first member, and apply the result of Art. 19 to a function of two variables, we find

$$d[f(x,y)] = du = \frac{du}{dx}dx + \frac{du}{dy}dy.$$

The differential of $f(x, y) = du$, which is obtained under the supposition that *both* the variables have changed their values, is called the *total differential* of the function.

85. If we have a function of three variables, as

$$u = f(x, y, z),$$

and suppose one of them, as z, to remain constant, and increments h and k to be attributed to the other two, the development of $f(x+h, y+k, z)$ will be of the same form as the development of $f(x+h, y+k)$; but u and all the differential coefficients will be functions of z.

If then an increment l be attributed to z, there will be four terms of the development of the form

$$u, \quad \frac{du}{dx}h, \quad \frac{du}{dy}k, \quad \frac{du}{dz}l.$$

If u were a function of four variables, as

$$u = f(x, y, z, s),$$

there would be five terms of the form

$$u, \quad \frac{du}{dx}h, \quad \frac{du}{dy}k, \quad \frac{du}{dz}l, \quad \frac{du}{ds}m;$$

and a new variable introduced into the function, would introduce a term containing the first power of its increment into the development.

If we transpose u into the first member, and make the same supposition as in the last article, we shall have

$$d[f(x, y, z)] = \frac{du}{dx}dx + \frac{du}{dy}dy + \frac{du}{dz}dz,$$

and, for like reasons,

$$d[f(x, y, z, s)] = \frac{du}{dx}dx + \frac{du}{dy}dy + \frac{du}{dz}dz + \frac{du}{ds}ds,$$

from which we may conclude that, *the total differential of a function of any number of variables is equal to the sum of the partial differentials.*

86. The rule demonstrated in the last article is alone sufficient for the differentiation of every algebraic function

1. Let $\quad u = x^2 + y^3 - z; \quad$ then

$$\frac{du}{dx}dx = 2x\,dx, \qquad \text{1st partial differential;}$$

$$\frac{du}{dy}dy = 3y^2dy, \quad \text{2d partial differential;}$$

$$\frac{du}{dz}dz = -dz, \quad \text{3d} \quad \text{"} \quad \text{"}$$

hence, $$du = 2x\,dx + 3y^2dy - dz.$$

2. Let $u = xy$; then,

$$\frac{du}{dx}dx = y\,dx,$$

$$\frac{du}{dy}dy = x\,dy:$$

hence, $$du = y\,dx + x\,dy.$$

3. Let $u = x^m y^n$; then,

$$\frac{du}{dx}dx = mx^{m-1}y^n dx,$$

$$\frac{du}{dy}dy = ny^{n-1}x^m dy: \qquad \text{hence,}$$

$$du = mx^{m-1}y^n dx + ny^{n-1}x^m dy = x^{m-1}y^{n-1}(mydx + nxdy)$$

4. Let $u = \frac{x}{y}$; then,

$$\frac{du}{dx}dx = \frac{dx}{y},$$

$$\frac{du}{dy}dy = -\frac{xdy}{y^2}$$

hence, $$du = \frac{y\,dx - x\,dy}{y^2}.$$

5. Let $u = \frac{ay}{\sqrt{x^2 + y^2}} = ay(x^2 + y^2)^{-\frac{1}{2}}$; then,

$$\frac{du}{dx}dx = -\frac{ayx\,dx}{(x^2 + y^2)^{\frac{3}{2}}},$$

$$\frac{du}{dy}dy = \frac{a\,dy}{(x^2 + y^2)^{\frac{1}{2}}} - \frac{ay^2\,dy}{(x^2 + y^2)^{\frac{3}{2}}},$$

hence, $$du = -\frac{ayx\,dx - ax^2\,dy}{(x^2 + y^2)^{\frac{3}{2}}}.$$

6. Let $u = xyzt$; then,

$$du = yztdx + xztdy + xytdz + xyzdt.$$

7. Let $u = z^y$; then,

$$\frac{du}{dy}dy = z^y lz\,dy \qquad \text{(Art. 55)},$$

$$\frac{du}{dx}dx = yz^{y-1}dz \qquad \text{(Art. 32)}$$

hence, $$du = z^y lz\,dy + yz^{y-1}dz.$$

Remark. In chapter II, the functions were supposed to depend on a common variable, and the differentials were obtained under this supposition. We now see that the dif ferentials are obtained in the same manner, when the functions are independent of each other, and unconnected with a common variable.

87. We have seen (Art. 39), that a function of a single variable has but one differential coefficient of the first order, one of the second, one of the third, &c.; while a

function of two variables has two differential coefficients of the first order, a function of three variables, three; a function of four variables, four; &c.

It is now proposed to find the successive differentials of a function of two variables, and also the successive differential coefficients.

We have already found

$$du = \frac{du}{dx}dx + \frac{du}{dy}dy.$$

But $$d(du) = d\left(\frac{du}{dx}dx + \frac{du}{dy}dy\right);$$

and since, $\frac{du}{dx}$ and $\frac{du}{dy}$ are functions of x and y, the differentials $\frac{du}{dx}dx$, $\frac{du}{dy}dy$, must each be differentiated with respect to both of the variables; dx and dy being supposed constant: hence,

$$d\left(\frac{du}{dx}dx\right) = \frac{d^2u}{dx^2}dx^2 + \frac{d^2u}{dxdy}dxdy,$$

and $$d\left(\frac{du}{dy}dy\right) = \frac{d^2u}{dy^2}dy^2 + \frac{d^2u}{dydx}dydx;$$

hence we have

$$d^2u = \frac{d^2u}{dx^2}dx^2 + 2\frac{d^2u}{dx\,dy}dx\,dy + \frac{d^2u}{dy^2}dy^2.$$

If we differentiate again, we have

$$d\left(\frac{d^2u}{dx^2}dx^2\right) = \frac{d^3u}{dx^3}dx^3 + \frac{d^3u}{dx^2}\frac{dx^2dy}{dy},$$

$$d\left(2\frac{d^2u}{dx\,dy}dx\,dy\right) = 2\frac{d^3u}{dx^2dy}dx^2dy + 2\frac{d^3u}{dx\,dy^2}dx\,dy^2,$$

$$d\left(\frac{d^2u}{dy^2}dy^2\right)=\frac{d^3u}{dy^2dx}dy^2dx+\frac{d^3u}{dy^3}dy^3;$$

and consequently,

$$d^3u=\frac{d^3u}{dx^3}dx^3+3\frac{d^3u}{dx^2dy}dx^2dy+3\frac{d^3u}{dx\,dy^2}dx\,dy^2+\frac{d^3u}{dy^3}dy^3.$$

It is very easy to find the subsequent differentials, by observing the analogy between the partial differentials and the terms of the development of a binomial.

We also see that, *a function of two variables has two partial differential coefficients of the first order, three of the second, four of the third,* &c.

88. There are several important results which may be deduced from the general development of the function of two variables (Art. 83).

1st. If we make $x=0$, and $y=0$, u and each of the differential coefficients will become constant, and we shall have

$$f(h,k)=u+\frac{1}{1}\left(\frac{du}{dx}h+\frac{du}{dy}k\right)$$

$$+\frac{1}{1.2}\left(\frac{d^2u}{dx^2}h^2+2\frac{d^2u}{dx\,dy}hk+\frac{d^2u}{dy^2}k^2\right)$$

$$+\ \&c.,$$

which is the development of any function of two variables in terms of their ascending powers, and coefficients which are dependent on the constants that enter the primitive function.

2d. If, in the general development, we make $y=0$, and $k=0$, we shall have

$$f(x+h) = u + \frac{du}{dx}\frac{h}{1} + \frac{d^2u}{dx^2}\frac{h^2}{1.2} + \frac{d^3u}{dx^3}\frac{h^3}{1.2.3} + \&c.,$$

which is the theorem of Taylor.

3d. If we make $y=0$, $k=0$, and $x=0$, we have

$$f(h) = u + \frac{du}{dx}\frac{h}{1} + \frac{d^2u}{dx^2}\frac{h^2}{1.2} + \frac{d^3u}{dx^3}\frac{h^3}{1.2.3} + \&c.,$$

or, $$f(h) = U + U'h + U''\frac{h^2}{1.2} + U'''\frac{h^3}{1.2.3} +, \&c.;$$

which is the theorum of Maclaurin.

Implicit Functions.

89. When the relation between a function and its variable is expressed by an equation of the form

$$y = f(x)$$

in which y is entirely disengaged from x, y has been called an *explicit*, or *expressed* function of x (Art. 6).
When y and x are connected together by an equation of the form

$$f(x,y) = 0,$$

y has been called an *implicit*, or *implied* function of x (Art. 6.)

It is plain, that in every equation of the form

$$f(x,y) = 0,$$

y must be a function of x, and x of y. For, if the equation were resolved with respect to either of them, the value found would be expressed in terms of the other variable and constant quantities.

90. If in the equation

$$u = f(x,y) = 0,$$

we suppose the variables x and y to change their values *in succession*, any change either in x or y, will produce a change in u: hence, u is a function of x and y when they vary in succession. The value, however, which u assumes, when x or y varies, will reduce to 0 when such a value is attributed to the other variable as will satisfy the equation

$$f(x,y) = 0.$$

We have from Art. 83,

$$f(x+h,\ y+k) - u = \frac{du}{dy}\frac{k}{1} + \text{terms containing } k^2,$$

$$\frac{du}{dx}h + \text{terms containing } h^2,$$

plus other terms containing kh, and the higher powers of h and k.

But, since y is a function of x, we have

$$k = Ph + P'h^2,$$

in which P is the differential coefficient of y regarded as a function of x. Substituting this value of k, and we have

$$(x+h,\ y+k) - u = \frac{du}{dy}Ph + \text{terms containing } h^2,$$

$$\frac{du}{dx}h + \text{terms containing } h^2,$$

plus other terms containing the higher powers of h.

But, in consequence of the relation between y and x, the first member of the equation will be constantly equal to 0. Hence, by the law of indeterminate coefficients (Alg., Art. 244),

$$\left(\frac{du}{dy}P + \frac{du}{dx}\right)h = 0\,;$$

hence,
$$P = \frac{dy}{dx} = -\frac{\frac{du}{dx}}{\frac{du}{dy}}.$$

Hence, *the differential coefficient of* y *regarded as a function of* x, *is equal to the ratio of the partial differential coefficients of* u *regarded as a function of* x, *and* u *regarded as a function of* y, *taken with a contrary sign.*

Let us take, as an example, the equation

$$f(x, y) = x^2 + y^2 - R^2 = u = 0\,;$$

then,
$$\frac{du}{dx} = 2x, \quad \text{and} \quad \frac{du}{dy} = 2y:$$

hence,
$$\frac{\frac{du}{dx}}{\frac{du}{dy}} = -\frac{x}{y} = \frac{dy}{dx}.$$

Although the differential coefficient of the first order is generally expressed in terms of x and y, yet y may be eliminated by means of the equation $f(x, y) = 0$, and the coefficient treated as a function of x alone. In the above equation we have

$$y = \sqrt{R^2 - x^2},$$

hence,
$$\frac{dy}{dx} = -\frac{x}{\sqrt{R^2 - x^2}}.$$

92. If it be required to find the second differential coefficient, we have merely to differentiate the first differential coefficient, regarded as a function of x, and divide the result by dx. Thus, if we designate the first differential coefficient by p, the second by q, the third by r, &c., we shall have

$$\frac{dp}{dx} = q, \qquad \frac{dq}{dx} = r, \quad \&c.$$

93. To find the second differential coefficient in the equation of the circle, we have

$$\frac{dy}{dx} = -\frac{x}{y},$$

$$d\left(\frac{dy}{dx}\right) = \frac{-ydx + xdy}{y^2}:$$

hence,
$$\frac{d^2y}{dx^2} = \frac{-y + x\dfrac{dy}{dx}}{y^2},$$

and by substituting for $\dfrac{dy}{dx}$ its value $-\dfrac{x}{y}$, we have

$$\frac{d^2y}{dx^2} = -\frac{x^2 + y^2}{y^3}.$$

1. Find the first differential coefficient of y, in the equation

$$y^2 - 2mxy + x^2 - a^2 = u = 0,$$

$$\frac{du}{dx} = -2my + 2x, \qquad \frac{du}{dy} = 2y - 2mx:$$

hence, $$\frac{dy}{dx} = -\left[\frac{-2my + 2x}{2y - 2mx}\right] = \frac{my - x}{y - mx}.$$

2. Find the first differential coefficient of y in the equation

$$y^2 + 2xy + x^2 - a^2 = 0.$$

$$\frac{dy}{dx} = -1.$$

3. Find the first and second differential coefficients of y, in the equation

$$y^3 - 3axy + x^3 = 0,$$

$$\frac{du}{dx} = 3x^2 - 3ay, \qquad \frac{du}{dy} = 3y^2 - 3ax,$$

hence, $$\frac{dy}{dx} = -\frac{3x^2 - 3ay}{3y^2 - 3ax} = \frac{ay - x^2}{y^2 - ax}.$$

For the second differential coefficient, we have

$$\frac{d^2y}{dx^2} = \frac{(y^2 - ax)\left(a\frac{dy}{dx} - 2x\right) - (ay - x^2)\left(2y\frac{dy}{dx} - a\right)}{(y^2 - ax)^2}:$$

or, by substituting for $\frac{dy}{dx}$ its value, and reducing,

$$\frac{d^2y}{dx^2} = -\frac{2xy^4 - 6ax^2y^2 + 2yx^4 + 2a^3xy}{(y^2 - ax)^3},$$

$$= -\frac{2xy(y^3 - 3axy + x^3) + 2a^3xy}{(y^2 - ax)^3}:$$

but from the given equation

$$y^3 - 3axy + x^3 = 0.$$

hence, $$\frac{d^2y}{dx^2} = -\frac{2a^3xy}{(y^2 - ax)^3}.$$

Differential Equations of Curves.

94. The Differential Calculus enables us to free an equation of its constants, and to find a new equation which shall only involve the variables and their differentials.

If, for example, we take the equation of a straight line

$$y = ax + b,$$

and differentiate it, we find

$$\frac{dy}{dx} = a,$$

and by differentiating again,

$$\frac{d^2y}{dx^2} = 0.$$

The last equation is entirely independent of the values of a and b, and hence, is equally applicable to every straight line which can be drawn in the plane of the co-ordinate axes. It is called, *the differential equation of lines of the first order.*

95. If we take the equation of the circle

$$x^2 + y^2 = R^2,$$

and differentiate it, we find

$$xdx + ydy = 0.$$

This equation is independent of the value of the radius R, and hence it belongs equally to every circle whose centre is at the origin of co-ordinates.

96. If the origin of co-ordinates be taken in the circumference, the equation of the circle (An. Geom. Bk. III, Prop. I, Sch. 3) is

$$y^2 = 2Rx - x^2;$$

from which we find

$$2R = \frac{y^2 + x^2}{x};$$

and by differentiating,

$$0 = \frac{x(2ydy + 2xdx) - (y^2 + x^2)dx}{x^2},$$

or by reducing

$$(x^2 - y^2)dx + 2xy\,dy = 0,$$

which is the differential equation of the circle when the origin of co-ordinates is in the circumference.

The last equation may be found in another manner.

If we differentiate the equation of the circle,

$$y^2 = 2Rx - x^2,$$

we have, after dividing by 2

$$ydy = Rdx - xdx;$$

hence, $$R = \frac{ydy + xdx}{dx}.$$

If this value of R be substituted in the equation of the circle, we have

$$(x^2 - y^2)dx + 2xy\,dy = 0;$$

the same differential equation as found by the first method.

97. If we take the general equation of lines of the second order (An. Geom. Bk. VI. Prop. XII, Sch. 3),

$$y^2 = mx + nx^2,$$

and differentiate it, we find

$$2ydy = mdx + 2nxdx\,;$$

differentiating again, regarding dx as constant, we have, after dividing by 2,

$$dy^2 + yd^2y = ndx^2.$$

Eliminating m and n from the three equations, we obtain

$$y^2dx^2 + x^2dy^2 - 2xy\,dx\,dy + yx^2d^2y = 0,$$

which is the general differential equation of lines of the second order.

98. In order to free an equation of its constants, it will be necessary to differentiate it as many times as there are constants to be eliminated. For, two equations are necessary to eliminate a single constant, three to eliminate two constants, four to eliminate three constants, &c.: hence, one constant may be eliminated from the given equation and the first differential equation; two from the given equation and the first and second differential equations, &c.

99. The differential equation which is obtained after the constants are eliminated, belongs to a *species* or *order of lines*, of which the given equation represents one of th species.

Thus, the differential equation (Art. 94),

$$\frac{d^2y}{dx^2} = 0,$$

belongs to an order or species of lines of which the equation

$$y = ax + b,$$

represents a single one, for given values of a and b.

The equation of a parabola is

$$y^2 = 2px,$$

and the differential equation of the species is

$$2xdy - ydx = 0, \qquad \text{or} \qquad dy^2 + yd^2y = 0.$$

100. The differential equation of a species, expresses the law by which the variable co-ordinates change their values; and this equation ought, therefore, to be independent of the constants which determine the *magnitude*, and not the *nature* of the curve.

101. The terms of an equation may be freed from their exponents, by differentiating the equation and then combining the differential and given equations.

Suppose, for example,

$$P^n = Q,$$

P and Q being any functions of x and y.

By differentiating, we obtain

$$nP^{n-1}dP = dQ:$$

by multiplying both members by P, we have

$$nP^n dP = PdQ,$$

and by substituting for P^n its value,

$$nQdP = PdQ.$$

The same result might also have been obtained by taking the logarithms of both members of the equation

$$P^n = Q.$$

For, we have

$$nlP = lQ,$$

and (Art. 57).

$$n\frac{dP}{P} = \frac{dQ}{Q};$$

hence,

$$nQdP = PdQ.$$

Of Vanishing Fractions, or those which take the form $\frac{0}{0}$.

102. It has been shown in (Alg. Art. 111), that $\frac{0}{0}$, though a symbol of an undetermined quantity, may, under particular suppositions, become equal to 0, to a finite quantity, or to infinity.

This symbol arises from the presence of a common factor in the numerator and denominator, which, becoming 0 for a particular value of the variable, reduces the fraction to the form $\frac{0}{0}$.

If we have, for example, a fraction of the form

$$\frac{P(x-a)^m}{Q(x-a)^n},$$

in which P and Q are functions of x, which do not reduce to 0, for $x = a$, we have

$$\frac{P(x-a)^m}{Q(x-a)^n} = \frac{0}{0}.$$

The value of this fraction will, however, be 0, finite or infinite, according as

$$m > n, \qquad m = n, \qquad m < n,$$

for under these suppositions, respectively, it takes the form

$$\frac{P(x-a)^{m-n}}{Q}, \qquad \frac{P}{Q}, \qquad \frac{P}{Q(x-a)^{n-m}}.$$

Let the numerator of the proposed fraction be designated by X, and the denominator by X', and let us suppose an arbitrary increment h to be given to x. The numerator and denominator will then become a function of $x+h$, and we shall have from the theorem of Taylor

$$\frac{X + \frac{dX}{dx}\frac{h}{1} + \frac{d^2X}{dx^2}\frac{h^2}{1.2} + \frac{d^3X}{dx^3}\frac{h^3}{1.2.3} + \&c.,}{X' + \frac{dX'}{dx}\frac{h}{1} + \frac{d^2X'}{dx^2}\frac{h^2}{1.2} + \frac{d^3X'}{dx^3}\frac{h^3}{1.2.3} + \&c.}$$

If the value of $x = a$, reduces to 0 the differential coefficients in the numerator as far as the mth order, and those of the denominator as far as the nth order, the value of the fraction will become,

$$\frac{\frac{d^mX}{dx^m}\frac{h^m}{1.2.3.4\ldots.m} + \&c.,}{\frac{d^nX'}{dx^n}\frac{h^n}{1.2.3.4\ldots.n} + \&c.}$$

If we make $h = 0$, the value of the fraction will become 0, finite, or infinite according as

$$m > n, \qquad m = n, \qquad m < n,$$

and hence, if the value $x = a$, reduces to 0 the same number of differential coefficients in the numerator and denominator, the value of the fraction will be finite and equal to the ratio of the first differential coefficients which do not reduce to 0.

103. Let us now illustrate this theory by examples.

1. If in the fraction

$$\frac{1-x^n}{1-x},$$

we make $x=1$, we have $\frac{0}{0}$. But

$$\frac{dX}{dx}=-nx^{n-1}, \qquad \frac{dX'}{dx}=-1;$$

in which, if we make $x=1$, we have

$$\frac{dX}{dx}=-n, \quad \text{and} \quad \frac{dX'}{dx}=-1,$$

hence,

$$\frac{\frac{dX}{dx}}{\frac{dX'}{dx}}=n,$$

therefore, the value of the fraction when $x=1$, is $+n$.

2. Find the value of the fraction

$$\frac{ax^2-2acx+ac^2}{bx^2-2bcx+bc^2}, \quad \text{when } x=c,$$

$$\frac{dX}{dx}=2ax-2ac, \qquad \frac{dX'}{dx}=2bx-2bc,$$

both of which become 0, when $x=c$. Differentiating again, we have

$$\frac{d^2X}{dx^2}=2a, \qquad \frac{d^2X'}{dx^2}=2b;$$

hence, the true value of the fraction when $x=c$ is $\frac{a}{b}$.

3. Find the value of the fraction

$$\frac{x^3 - ax^2 - a^2x + a^3}{x^2 - a^2}, \text{ when } x = a.$$

Ans. 0.

4. Find the value of the fraction

$$\frac{ax - x^2}{a^4 - 2a^3x + 2ax^3 - x^4}, \text{ when } x = a.$$

Ans. ∞.

5. Find the value of

$$\frac{a^x - b^x}{x}, \text{ when } x = 0.$$

Ans. $la - lb$.

6. What is the value of the fraction

$$\frac{1 - \sin x + \cos x}{\sin x + \cos x - 1}, \text{ when } x = 90^\circ.$$

Ans. 1.

7. What is the value of the fraction

$$\frac{a - x - ala + alx}{a - \sqrt{2ax - x^2}}, \text{ when } x = a.$$

Ans. -1.

8. What is the value of the fraction

$$\frac{x^x - x}{1 - x + lx}, \text{ when } x = 1.$$

Ans. -2.

9. What is the value of the fraction

$$\frac{a^n - x^n}{la - lx}, \text{ when } x = a.$$

Ans. na^n.

104. It has been remarked (Art. 47), that the theorem of Taylor does not apply to the case in which a particular value attributed to x, renders any differential coefficient of the function infinite. Such functions are of the form

$$\frac{(x^2-a^2)^m}{(x-a)^n},$$

in which m and n are fractional.

In functions of this form we substitute for x, $a+h$, which gives a second state of the function. We then divide the numerator and denominator by h raised to a power denoted by the smallest exponent of h, after which we make $h=0$, and find the ratio of the terms of the fraction.

When we place $a+h$ for x, we have, in arranging according to the ascending powers of h,

$$\frac{F(a+h)}{F'(a+h)}=\frac{Ah^a+Bh^b+Ch^c+\&c.,}{A'h^{a'}+Bh^{b'}+Ch^{c'}+\&c.}$$

Now there are three cases, viz.: when

$$a>a', \qquad a=a', \qquad a<a'.$$

In the first case the value of the fraction will be 0; in the second, a finite quantity; and in the third it will be infinite.

105. In substituting $a+h$ for x, in the fraction

$$\frac{(x^2-a^2)^{\frac{3}{2}}}{(x-a)^{\frac{3}{2}}},$$

we have
$$\frac{(2ah+h^2)^{\frac{3}{2}}}{h^{\frac{3}{2}}}=(2a+h)^{\frac{3}{2}},$$

and by making $h=0$, which renders $x=a$ the value of the fraction becomes
$$(2a)^{\frac{3}{2}}.$$

2. Required the value of the fraction
$$\frac{(x^2-3ax+2a^2)^{\frac{2}{3}}}{(x^3-a^3)^{\frac{1}{2}}} \quad \text{when} \quad x=a.$$

Substituting $a+h$ for x, we have
$$\frac{h^{\frac{2}{3}}(-a+h)^{\frac{2}{3}}}{h^{\frac{1}{2}}(3a^2+3ah+h^2)^{\frac{1}{2}}}=\frac{h^{\frac{1}{6}}(-a+h)^{\frac{2}{3}}}{(3a^2+3ah+h^2)^{\frac{1}{2}}},$$

which is equal to 0, when $h=0$.

106. *Remark.* The last method of finding the value of a vanishing fraction, may frequently be employed advantageously, even when the value can be found by the theorem of Taylor.

107. There are several forms of indetermination under which a function may appear, but they can all be reduced to the form $\frac{0}{0}$.

1st. Suppose the numerator and denominator of the fraction
$$\frac{X}{X'},$$

to become infinite by the supposition of $x=a$. The fraction can be placed under the form

$$\frac{\frac{1}{X'}}{\frac{1}{X}},$$

which reduces to $\frac{0}{0}$, when X and X' are infinite.

2d. We may have the product of two factors, one of which becomes 0 and the other infinite, when a particular value is given to the variable.

In the product PQ, let us suppose that $x=a$, makes $P=0$ and $Q=\infty$. We would then write the product under the form,

$$PQ=\frac{P}{\frac{1}{Q}}$$

which becomes $\frac{0}{0}$ when $x=a$.

108. Let us take, as an example, the function

$$(1-x)\operatorname{tang}\frac{1}{2}\pi x;$$

in which π designates 180°.

If we make $x=1$, the first factor becomes 0, and the second infinite. But

$$\operatorname{tang}\frac{1}{2}\pi x=\frac{1}{\cot\frac{1}{2}\pi x};$$

hence, $$(1-x)\operatorname{tang}\frac{1}{2}\pi x=\frac{1-x}{\cot\frac{1}{2}\pi x},$$

the value of which is $\frac{2}{\pi}$ when $x=1$.

CHAPTER V.

Of the Maxima and Minima of a Function of a Single Variable.

109. If we have

$$u = f(x),$$

the value of the function u may be changed in two ways: first, by increasing the variable x; and secondly, by diminishing it.

If we designate by u' the *first* value which u assumes when x is increased, and by u'' the *first* value which u assumes when x is diminished, we shall have three consecutive values of the function

$$u', \quad u, \quad u''.$$

Now, when u is greater than both u' and u'', u is said to be a *maximum:* and when u is less than both u' and u'', it is said to be a *minimum.*

Hence, *the maximum value of a variable function is greater than the value which immediately precedes, or the value that immediately follows: and the minimum value of a variable function is less than the value which immediately precedes or the value that immediately follows.*

110. Let us now determine the analytical conditions which characterize the maximum and minimum values of a variable function.

If in the function

$$u = f(x),$$

the variable x be first increased by h, and then diminished by h, we shall have (Art. 44),

$$u' = f(x+h) = u + \frac{du}{dx}\frac{h}{1} + \frac{d^2u}{dx^2}\frac{h^2}{1.2} + \frac{d^3u}{dx^3}\frac{h^3}{1.2.3} + \&c.,$$

$$u'' = f(x-h) = u - \frac{du}{dx}\frac{h}{1} + \frac{d^2u}{dx^2}\frac{h^2}{1.2} - \frac{d^3u}{dx^3}\frac{h^3}{1.2.3} + \&c.;$$

and consequently,

$$u' - u = \frac{du}{dx}\frac{h}{1} + \frac{d^2u}{dx^2}\frac{h^2}{1.2} + \frac{d^3u}{dx^3}\frac{h^3}{1.2.3} + \&c.,$$

$$u'' - u = -\frac{du}{dx}\frac{h}{1} + \frac{d^2u}{dx^2}\frac{h^2}{1.2} - \frac{d^3u}{dx^3}\frac{h^3}{1.2.3} + \&c.$$

Now, if u has a maximum value, it will be *greater* than u' or u''; and hence, $u' - u$ and $u'' - u$ will both be negative. If u is a minimum, it will be *less* than u' or u'', and hence, $u' - u$ and $u'' - u$ will both be positive.

Hence, in order that u may have a maximum or a minimum value, the signs of the two developments must be both minus or both plus.

But since the terms involving the first power of h, in the two developments, have contrary signs, and since so small a value may be assigned to h as to make the first term in each development greater than the sum of all t other terms (Art. 44), it follows that u can have neitl a maximum nor a minimum, unless

$$\frac{du}{dx} = 0;$$

and the roots of this equation will give all the values of x which can render the function u either a maximum or a minimum.

Having made the first differential coefficient equal to 0, the signs of the developments will depend on the sign of second differential coefficient.

But since the signs of the first members of the equations, and consequently of the developments, are both negative when u is a maximum, and both positive when u is a minimum, it follows that the second differential coefficient will be negative when the function is a maximum, and positive when it is a minimum. Hence, the roots of the equation

$$\frac{du}{dx}=0,$$

being substituted in the second differential coefficient, will render it negative in case of a maximum, and positive in case of a minimum; and since there may be more than one value of x which will satisfy these conditions, it follows that there may be more than one maximum or one minimum.

But if the roots of the equation

$$\frac{du}{dx}=0,$$

reduce the second differential coefficient to 0, the signs of the developments will depend on the signs of the terms which involve the third differential coefficient; and these signs being different, there can neither be a maximum nor a minimum, unless the values of x also reduce the third differential coefficient to 0. When this is the case, substitute the roots of the equation

$$\frac{du}{dx}=0,$$

in the fourth differential coefficient; if it becomes negative there will be a maximum, if positive a minimum. If the values of x reduce the fourth differential coefficient to 0, the following differential coefficient must be examined. Hence, in order to find the values of x which will render the proposed function a maximum or a minimum.

1st. *Find the roots of the equation*

$$\frac{du}{dx}=0.$$

2d. *Substitute these roots in the succeeding differential coefficients, until one is found which does not reduce to* 0. *Then, if the differential coefficient so found be of an odd order, the values of* x *will not render the function either a maximum or a minimum. But if it be of an even order, and negative, the function will be a maximum; if positive, a minimum.*

111. *Remark.* Before applying the preceding rules to examples, it may be well to remark, that if a variable function is multiplied or divided by a constant quantity, the same values of the variable which render the function a maximum or a minimum, will also make the product or quotient a maximum or a minimum, and hence the constant will not affect the conditions of maximum or minimum.

2. Any value of the variable which will render the function a maximum or a minimum, will also render any root or power a maximum or a minimum; and hence, if a function is under a radical, the radical may be omitted.

EXAMPLES.

1. To find the value of x which will render y a maximum or a minimum in the equation of the circle

$$y^2 + x^2 = R^2,$$

$$\frac{dy}{dx} = -\frac{x}{y},$$

making $-\frac{x}{y} = 0$, gives $x = 0$.

The second differential coefficient is

$$\frac{d^2y}{dx^2} = -\frac{x^2 + y^2}{y^3},$$

and since making $x = 0$, gives $y = R$, we have

$$\frac{d^2y}{dx^2} = -\frac{1}{R}$$

which being negative, the value of $x = 0$ renders y a maximum.

2. Find the values of x which will render y a maximum or a minimum in the equation,

$$y = a - bx + x^2,$$

differentiating, we find

$$\frac{dy}{dx} = -b + 2x, \quad \text{and} \quad \frac{d^2y}{dx^2} = 2,$$

making, $-b + 2x = 0$, gives $x = \frac{b}{2}$;

and since the second differential coefficient is positive, this value of x will render y a minimum. The minimum value of y is found by substituting the value of x, in the primitive equation. It is

$$y = a - \frac{b^2}{4}.$$

3. Find the value of x which will render the function

$$u = a^4 + b^3x - c^2x^2,$$

a maximum or a minimum,

$$\frac{du}{dx} = b^3 - 2c^2x, \qquad \text{hence} \qquad x = \frac{b^3}{2c^2};$$

and,

$$\frac{d^2u}{dx^2} = -2c^2:$$

hence, the function is a maximum, and the maximum value is

$$u = a^4 + \frac{b^6}{4c^2}.$$

4. Let us take the function

$$u = 3a^2x^3 - b^4x + c^5,$$

we find $\frac{du}{dx} = 9a^2x^2 - b^4$, and $x = \pm\frac{b^2}{3a}$

The second differential coefficient is

$$\frac{d^2u}{dx^2} = 18a^2x$$

Substituting the plus root of x, we have

$$\frac{d^2u}{dx^2} = +6ab^2,$$

which gives a minimum, and substituting the negative root, we have

$$\frac{d^2u}{dx^2} = -6ab^2,$$

which gives a maximum.

The minimum value of the function is,

$$u = c^5 - \frac{2b^6}{9a};$$

and the maximum value

$$u = c^5 + \frac{2b^6}{9a}.$$

112. *Remark.* It frequently happens that the value of the first differential coefficient may be decomposed into two factors, X and X', each containing x, and one of them, X for example, reducing to 0 for that value of x, which renders the function a maximum or a minimum. When the differential coefficient of the first order takes this form, the general method of finding the second differential coefficient may be much simplified. For, if

$$\frac{du}{dx} = XX',$$

we shall have

$$\frac{d^2u}{dx^2} = \frac{X'dX}{dx} + \frac{XdX'}{dx}.$$

But by hypothesis X reduces to 0 for that value of x which renders the function u a maximum or a minimum:

hence, $$\frac{d^2u}{dx^2} = \frac{X'dX}{dx};$$

from which we obtain the following rule for finding the second differential coefficient.

Differentiate that factor of the first differential coefficient which reduces to 0, *multiply it by the other factor, and divide the product by* dx.

5. To divide a quantity into two such parts that the mth power of one of the parts multiplied by the nth power of the other shall be a maximum or a minimum.

Designate the given quantity by a and one of the parts by x, then will $a-x$ represent the other part. Let the product of their powers be designated by u; we shall then have

$$u = x^m(a-x)^n,$$

whence, $$\frac{du}{dx} = mx^{m-1}(a-x)^n - nx^m(a-x)^{n-1},$$

$$= (ma - mx - nx)x^{m-1}(a-x)^{n-1},$$

and by placing each of the factors equal to 0, we have

$$x = \frac{ma}{m+n}, \qquad x = 0, \qquad x = a.$$

The second differential coefficient corresponding to the first of these values, found by the method just explained, is

$$\frac{d^2u}{dx^2} = -(m+n)x^{m-1}(a-x)^{n-1};$$

and substituting for x its value, it becomes

$$-\frac{m^{m-1}n^{n-1}a^{m+n-2}}{(m+n)^{m+n-3}}:$$

hence, this value of x renders the product a maximum. The two other values of x satisfy the *equation* of the problem, but do not satisfy the enunciation, since they are not *parts* of the given quantity a.

Remark. If m and n are each equal to unity, the quantity will be divided into equal parts.

6. To determine the conditions which will render y a maximum or a minimum in the equation

$$y^2 - 2mxy + x^2 - a^2 = 0.$$

The first differential coefficient is

$$\frac{dy}{dx} = \frac{my - x}{y - mx};$$

hence, $my - x = 0$, or $y = \frac{x}{m}$.

Substituting this value of y in the given equation, we find

$$x = \frac{ma}{\sqrt{1 - m^2}};$$

and the value of y corresponding to this value of x is

$$y = \frac{a}{\sqrt{1 - m^2}}.$$

To determine whether y is a maximum or a minimum, let us pass to the second differential coefficient. We have

$$\frac{dy}{dx} = (my - x)(y - mx)^{-1};$$

hence,
$$\frac{d^2y}{dx^2} = \frac{\left(m\frac{dy}{dx} - 1\right)}{y - mx}:$$

and since $\frac{dy}{dx}=0$, we have

$$\frac{d^2y}{dx^2}=-\frac{1}{y-mx},$$

and by substituting for y and x their values, we have

$$\frac{d^2y}{dx^2}=-\frac{1}{a\sqrt{1-m^2}};$$

hence, y is a maximum.

7. To find the maximum rectangle which can be inscribed in a given triangle.

Let b denote the base of the triangle, h the altitude, y the base of the rectangle, and x the altitude. Then,

$$u=xy=\text{ the area of the rectangle.}$$

But $$b : h :: y : h-x:$$

hence, $$y=\frac{bh-bx}{h},$$

and consequently,

$$u=\frac{bhx-bx^2}{h}=\frac{b}{h}(hx-x^2).$$

and omitting the constant factor,

$$\frac{du}{dx}=h-2x, \quad \text{or} \quad x=\frac{h}{2};$$

hence, the altitude of the rectangle is equal to half the altitude of the triangle: and since

$$\frac{d^2u}{dx^2}=-2,$$

the area is a maximum.

8. What is the altitude of a cylinder inscribed in a given cone, when the solidity of the cylinder is a maximum?

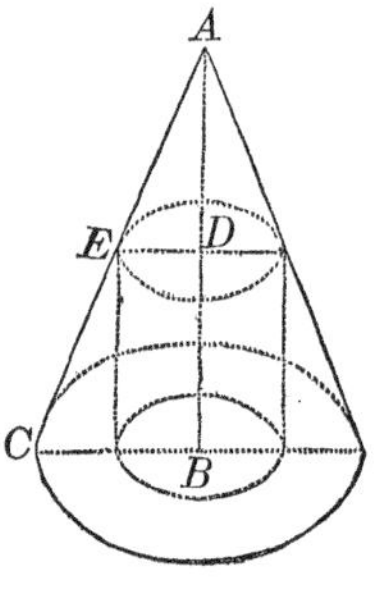

Suppose the cylinder to be inscribed, as in the figure, and let $AB = a$, $BC = b$, $AD = x$, $ED = y$: then, $BD = a - x =$ altitude of the cylinder, and $\pi y^2(a - x) =$ solidity $= v$. (1)

From the similar triangles AED and ACB, we have

$$x : y :: a : b; \quad \text{whence} \quad y = \frac{bx}{a}.$$

Substituting this value in equation (1), and we have

$$v = \frac{\pi b^2}{a^2}x^2(a - x).$$

Omitting the constant factor $\frac{\pi b^2}{a^2}$, we may write

$$u = x^2(a - x);$$

for the conditions which will make u a maximum will also make v a maximum (Art. 111).

By differentiating, we have

$$\frac{du}{dx} = 2ax - 3x^2, \text{ and } \frac{d^2u}{ax^2} = 2a - 6x.$$

Placing $\qquad 2ax - 3x^2 = 0,$

we have $\qquad x = 0, \text{ and } x = \frac{2}{3}a.$

Hence the altitude of the maximum cylinder is one-third the altitude of the cone

9. What are the sides of the maximum rectangle inscribed in a given circle?

Ans. Each equal to $R\sqrt{2}$.

10. A cylindrical vessel is to contain a given quantity of water. Required the relation between the diameter of the base and the altitude in order that the interior surface may be a minimum.

Ans. Altitude = radius of base.

11. To find the altitude of a cone inscribed in a given sphere, which shall render the convex surface of the cone a maximum.

Ans. Altitude $= \frac{4}{3}R$.

12. To find the maximum right-angled triangle which can be described on a given line.

Ans. When the two sides are equal.

13. What is the length of the axis of the maximum parabola that can be cut from a given right cone?

Ans. Three-fourths the side of the cone.

14. To find the least triangle which can be formed by the radii produced, and a tangent line to the quadrant of a given circle.

Ans. When the point of contact is at the middle of the arc.

15. What is the altitude of the maximum cylinder which can be inscribed in a given paraboloid?

Ans. Half the axis of the paraboloid

CHAPTER VI.

Application of the Differential Calculus to the Theory of Curves.

113. Every relation between a function y and a variable x, expressed by the equation

$$y = f(x),$$

will subsist between the ordinate and abscissa of a curve

Let A be the origin of the rectangular axes; then if in the equation

$$y = f(x),$$

we make $x = 0$, we have

$$y = \text{a constant}:$$

lay off AB equal to this constant. Then attribute values to x, from 0 to any limit, as well negative as positive, and find from the equation

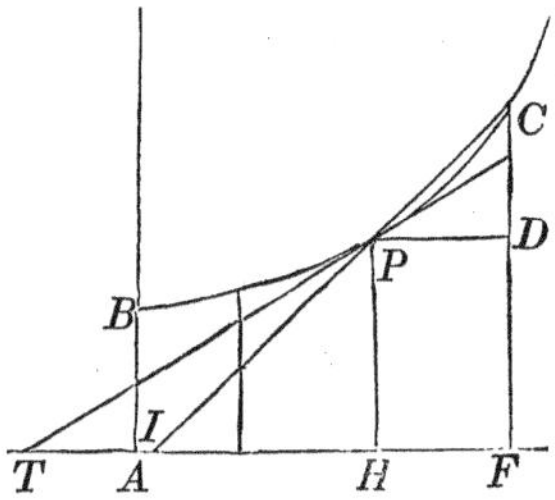

$$y = f(x),$$

the corresponding values of y. Conceive the values of x to be laid off on the axis of abscissas, and the values of y on the corresponding ordinates. The curve described through the extremities of the ordinates will have for its equation

$$y = f(x). \quad (1)$$

Let x represent any abscissa, AH for example, and y the corresponding ordinate HP.

If now we give to x any arbitrary increment h, and make $HF = h$, the value of y will become equal to FC, which we will designate by y'. Through P draw the secant CPI and the tangent TP.

Now, $y' - y = CF - PH = CD,$

and $\frac{y'-y}{h} = \frac{CD}{PD}$ = tangent of the angle CPD.

But, by similar triangles $\frac{CD}{PD} = \frac{PH}{IH}$.

Now, the limiting ratio of the increment of the variable to that of the function, is that ratio which is independent of the value of h, and is obtained by making h equal to 0 in the expression for the ratio of the increments (Art. 15.)

It is evident that as h diminishes, the point C will approach the point P, the point I will approach T, and the secant IC will approach the tangent TP; and when h becomes equal to 0, the secant IC will coincide with the tangent TP. For every position of C we shall have $\frac{CD}{PD} = \frac{PH}{TH}$ = tangent CPD = tangent CIH; and when C coincides with P, $\frac{CD}{PD} = \frac{PH}{TH}$ = tangent $PTH = \frac{dy}{dx}$; that is, *the limiting ratio, or first differential coefficient, is equal to the tangent of the angle which the tangent line makes with the axis of abscissas.*

Of Tangents and Normals.

114. Having found the value of

$$\frac{dy}{dx}$$

we will now proceed to find the value of the subtangent, tangent, subnormal, and normal.

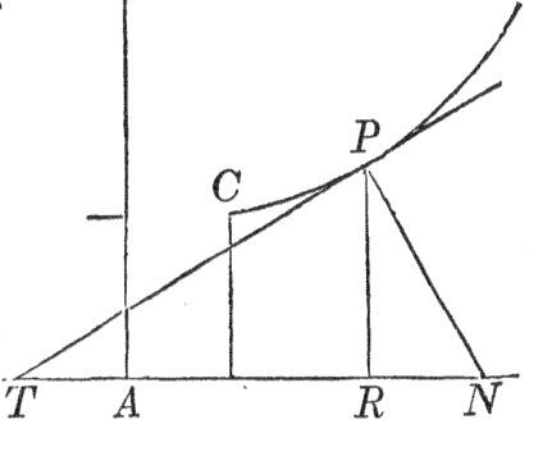

We have (Trig. Th. II),

$$1 : TR :: \text{tang}\, T : RP;$$

that is,

$$1 : TR :: \frac{dy}{dx} : y\cdot$$

hence,

$$TR = y\frac{dx}{dy} = \text{sub-tangent}.$$

115. The tangent TP is equal to the square root of the sum of the squares of TR and RP; hence,

$$TP = y\sqrt{1 + \frac{dx^2}{dy^2}} = \text{tangent}.$$

116. From the similar triangles TPR, RPN, we have

$$TR : PR :: PR : RN,$$

hence,

$$y\frac{dx}{dy} : y :: y : RN,$$

consequently,

$$RN = y\frac{dy}{dx} = \text{sub-normal}.$$

117. The normal PN is equal to the square root of the sum of the squares of PR and RN; hence,

$$PN = y\sqrt{1 + \frac{dy^2}{dx^2}} = \text{normal}.$$

118. Let it be now required to apply these formulas to lines of the second order, of which the general equation (An. Geom. Bk. VI, Prop. XII, Sch. 3), is,

$$y^2 = mx + nx^2.$$

Differentiating, we have

$$\frac{dy}{dx} = \frac{m + 2nx}{2y} = \frac{m + 2nx}{2\sqrt{mx + nx^2}};$$

substituting this value, we find

$$\text{sub-tangent } TR = y\frac{dx}{dy} = \frac{2(mx+nx^2)}{m+2nx},$$

$$TP = y\sqrt{1+\frac{dx^2}{dy^2}} = \sqrt{mx+nx^2+4\left[\frac{mx+nx^2}{m+2nx}\right]^2},$$

$$\text{sub-normal } RN = y\frac{dy}{dx} = \frac{m+2nx}{2},$$

$$PN = y\sqrt{1+\frac{dy^2}{dx^2}} = \sqrt{mx+nx^2+\frac{1}{4}(m+2nx)^2}.$$

By attributing proper values to m and n, the above formulas will become applicable to each of the conic sections. In the case of the parabola, $n=0$, and we have

$$TR = 2x, \qquad TP = \sqrt{mx+4x^2},$$

$$RN = \frac{m}{2}, \qquad PN = \sqrt{mx+\frac{1}{4}m^2}.$$

119. It is often necessary to represent the tangent and normal lines by their equations. To determine these, in a general manner, it will be necessary first to consider the analytical conditions which render any two curves tangent to each other.

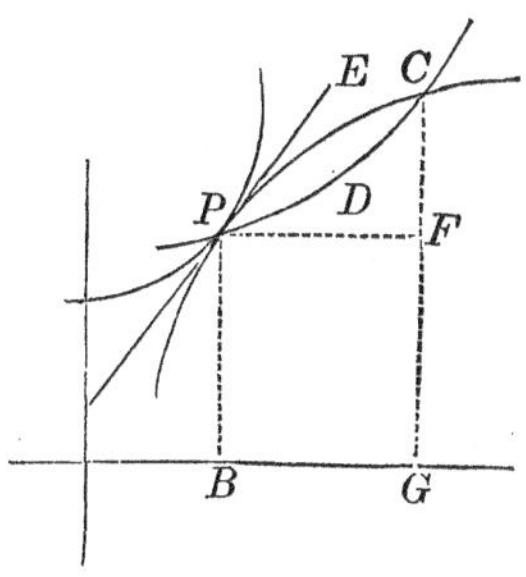

Let the two curves, PDC, PEC, intersect each other at P and C.

Designate the co-ordinates of the first curve by x and y, and the co-ordinates of the second by x', y'. Then, for the common point P, we shall have

$$x = x', \qquad y = y'.$$

If we represent BG, the increment of the abscissa, by h, we shall have, from the theorem of Taylor (Art. 44),

$$CG - PB = CF = \frac{dy}{dx}\frac{h}{1} + \frac{d^2y}{dx^2}\frac{h^2}{1.2} + \frac{d^3y}{dx^3}\frac{h^3}{1.2.3} + \&c.,$$

$$CG - PB = CF = \frac{dy'}{dx'}\frac{h}{1} + \frac{d^2y'}{dx'^2}\frac{h^2}{1.2} + \frac{d^3y'}{dx'^3}\frac{h^3}{1.2.3} + \&c.$$

hence, by placing the two members equal to each other, and, dividing by h, we have

$$\frac{dy}{dx} + \frac{d^2y}{dx^2}\frac{h}{1.2} + \&c., = \frac{dy'}{dx'} + \frac{d^2y'}{dx'^2}\frac{h}{1.2} + \&c.$$

If we now pass to the limit, by making $h = 0$, we shall have

$$\frac{dy}{dx} = \frac{dy'}{dx'};$$

in which case the point C will become consecutive with P, and the curve PEC tangent to the curve PDC. Hence, *two lines will be tangent to each other, when they have a common point, and the first differential coefficient of the one equal to the first differential coefficient of the other, for this point.*

120. The equation of a straight line is of the form

$$y = ax + b,$$

hence,

$$\frac{dy}{dx} = a.$$

But the equation of a straight line passing through a given point, of which the co-ordinates are x'', y'', is (An. Geom. Bk. II, Prop. IV),

$$y - y'' = a(x - x'').$$

But if the point whose co-ordinates are x'', y'', is required to be on a given curve, these co-ordinates must satisfy the equation of that curve. If the straight line is required to be tangent to the curve at this particular point, the first differential coefficient $\frac{dy}{dx}$, found from the equation of the curve, must take the particular value $\frac{dy''}{dx''}$; that is, we must have

$$\frac{dy}{dx} = \frac{dy''}{dx''},$$

and the equation of the line tangent at the point whose co-ordinates are x'', y'', will be

$$y - y'' = \frac{dy''}{dx''}(x - x'').$$

121. Let it be required, for example, to make the line tangent to the circumference of a circle at a point of which the co-ordinates are x'', y''.

The equation of the circle is $x^2 + y^2 = R^2$;

and, by differentiating, we have $\frac{dy}{dx} = -\frac{x}{y}$.

But if the straight line is to be tangent to the circle, at the point whose co-ordinates are x'', y'', we must have

$$\frac{dy''}{dx''} = \frac{dy}{dx} = -\frac{x}{y} = -\frac{x''}{y''};$$

and by substituting this value in the equation of the line, and recollecting that $x''^2 + y''^2 = R^2$, we have

$$yy'' + xx'' = R^2,$$

which is the equation of a tangent line to a circle.

122. A normal line is perpendicular to the tangent at

the point of contact, and since the equation of the tangent is of the form

$$y - y'' = \frac{dy''}{dx''}(x - x''),$$

the equation of the normal, at the point whose co-ordinates are x'', y'', will be of the form (An. Geom. Bk. II., Prop. VII., Sch. 2),

$$y - y'' = -\frac{dx''}{dy''}(x - x''),$$

If we take the equation of any curve, and find the value of $\frac{dx''}{dy''}$ for the particular point whose co-ordinates are x'', y'', and then substitute that value in the above equation, we shall have the equation of the normal passing through this point.

The equation of the normal in the circle will take the form

$$y = \frac{y''}{x''}x.$$

123. To find the equation of a tangent line to an ellipse at a point of which the co-ordinates are x'', y'', we have

$$A^2y''^2 + B^2x''^2 = A^2B^2.$$

By differentiating, we have

$$\frac{dy''}{dx''} = -\frac{B^2x''}{A^2y''};$$

hence, we have

$$y - y'' = -\frac{B^2x''}{A^2y''}(x - x''), \text{ or } A^2yy'' + B^2xx'' = A^2B^2;$$

and for the normal $\quad y - y'' = \frac{A^2y''}{B^2x''}(x - x'')$.

124. To find the equation of a tangent to lines of the second order, of which the equation for a particular point (An. Geom. Bk. VI, Prop. XII, Sch. 3) is

$$y''^2 = mx'' + nx''^2.$$

By differentiating, we have

$$\frac{dy''}{dx''} = \frac{m + 2nx''}{2y''};$$

hence, the equation of the tangent to a line of the second order is

$$y - y'' = \frac{m + 2nx''}{2y''}(x - x''),$$

and the equation of the normal

$$y - y'' = -\frac{2y''}{m + 2nx''}(x - x'').$$

Of Asymptotes of Curves.

125. An asymptote of a curve is a line which continually approaches the curve, and becomes tangent to it at an infinite distance from the origin of co-ordinates.

Let AX and AY be the co-ordinate axes, and

$$y - y'' = \frac{dy''}{dx''}(x - x''),$$

the equation of any tangent line, as TP.

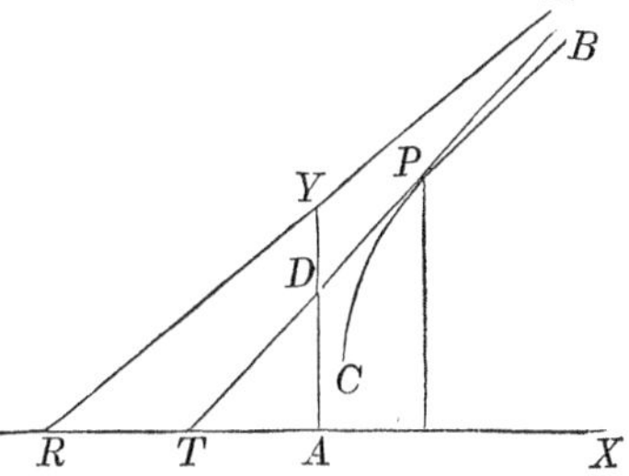

If in the equation of the tangent, we make in succession $y=0$, $x=0$, we shall find

$$x=AT=x''-y''\frac{dx''}{dy''}, \qquad y=AD=y''-x''\frac{dy''}{dx''}.$$

If the curve CPB has an asymptote RE, it is plain that the tangent PT will approach the asymptote RE, when the point of contact P, is moved along the curve from the origin of co-ordinates, and T and D will also approach the points R and Y, and will coincide with them when the co-ordinates of the point of tangency are infinite.

In order, therefore, to determine if a curve have asymptotes, we substitute in the values of AT and AD, the co-ordinates of the point which is at an infinite distance from the origin of co-ordinates. If either of the distances AT, AD, become finite, the curve will have an asymptote.

If both the values are finite, the asymptote will be inclined to both the co-ordinate axes: if one of the distances becomes finite and the other infinite, the asymptote will be parallel to one of the co-ordinate axes; and if they both become 0, the asymptote will pass through the origin of co-ordinates. In the last case, we shall know but one point of the asymptote, but its direction may be determined by finding the value of $\frac{dy}{dx}$, under the supposition that the co-ordinates are infinite.

126. Let us now examine the equation

$$y^2=mx+nx^2,$$

of lines of the second order, and see if these lines have asymptotes. We find

$$AT = x - \frac{2y^2}{m + 2nx} = \frac{-mx}{m + 2nx},$$

$$AD = y - \frac{mx + 2nx^2}{2y} = \frac{mx}{2\sqrt{mx + nx^2}};$$

which may be put under the forms

$$AT = \frac{-m}{\frac{m}{x} + 2n}, \qquad AD = \frac{m}{2\sqrt{\frac{m}{x} + n}},$$

and making $x = \infty$, we have

$$AR = -\frac{m}{2n}, \qquad \text{and} \qquad AE = \frac{m}{2\sqrt{n}}.$$

If now we make $n = 0$, the curve becomes a parabola, and both the limits, AR, AE, become infinite: hence, the parabola has no rectilinear asymptote.

If we make n negative, the curve becomes an ellipse, and AE becomes imaginary: hence, the ellipse has no asymptote.

But if we make n positive, the equation becomes that of the hyperbola, and both the values, AR, AE, become finite. If we substitute for n its value $\frac{B^2}{A^2}$, we shall have

$$AR = -A, \qquad \text{and} \qquad AE = \pm B.$$

Differentials of the Arcs and Areas of Segments of Curves.

127. It is plain, that the chord and arc of a curve will approach each other continually as the arc is diminished, and hence, we might conclude that the limit of their ratio is unity. But as several propositions depend on this relation between the arc and chord, we shall demonstrate it rigorously.

128. If we suppose the ordinate PR of the curve, POM to be a function of the abscissa, we shall have (Art. 19),

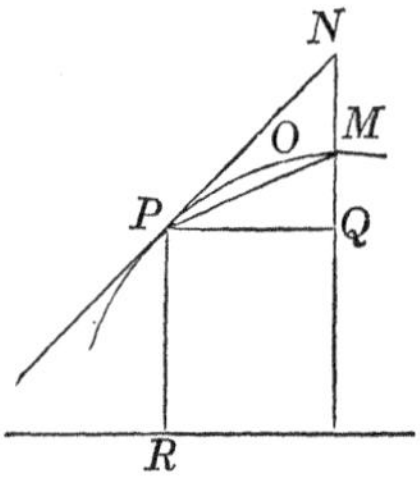

$$PQ = h, \qquad \text{and}$$

$$y' - y = MQ = (P + P'h)h;$$

in which
$$P = \frac{dy}{dx}.$$

Hence, $$PM = \sqrt{h^2 + (P + P'h)^2 h^2} = h\sqrt{1 + (P + P'h)^2}.$$

We also have $$NQ = Ph;$$

hence, $$PN = \sqrt{h^2 + P^2h^2} = h\sqrt{1 + P^2},$$

$$NM = NQ - MQ = -P'h^2.$$

hence, we have

$$\frac{PN + MN}{PM} = \frac{h\sqrt{1 + P^2} - P'h^2}{h\sqrt{1 + (P + P'h)^2}} = \frac{\sqrt{1 + P^2} - P'h}{\sqrt{1 + (P + P'h)^2}};$$

of which the limit, by making $h=0$, is

$$\frac{\sqrt{1+P^2}}{\sqrt{1+P^2}}=1.$$

But the arc POM can never be less than the chord PM, nor greater than the broken line PNM which contains it; hence, the limit of the ratio

$$\frac{POM}{PM}=1;$$

and consequently, *the differential of the arc is equal to the differential of the chord.* If we designate the arc by z, PM will be represented by $z'-z$, and we shall have

$$\frac{z'-z}{h}=\frac{POM}{PM}\times\frac{PM}{PQ}=\frac{POM}{PM}\times\frac{h}{h}\sqrt{1+(P+P'h)^2};$$

and, by passing to the limiting ratio,

$$\frac{dz}{dx}=\sqrt{1+P^2}=\sqrt{1+\frac{dy^2}{dx^2}}.$$

or $$dz=\sqrt{dx^2+dy^2};$$

that is, *the differential of the arc of a curve, at any point, is equal to the square root of the sum of the squares of the differentials of the co-ordinates.*

129. To determine the differential of the arc of a circle of which the equation is

$$x^2+y^2=R^2,$$

we have $xdx+ydy=0$, or $dy=-\dfrac{xdx}{y}$;

whence, $dz=\sqrt{dx^2+\dfrac{x^2dx^2}{y^2}}=\dfrac{dx}{y}\sqrt{x^2+y^2},$

$$=\frac{Rdx}{y}=\pm\frac{Rdx}{\sqrt{R^2-x^2}},$$

the same as determined in (Art 71). The plus sign is to be used when the abscissa x and the arc are increasing functions of each other, and the minus sign when they are decreasing functions (Art. 31).

130. Let BCD be any segment of a curve, and let it be required to find the differential of its area.

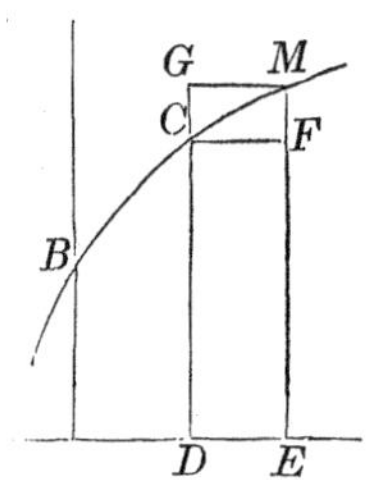

The two rectangles $DCFE$, $DGME$, having the same base DE, are to each other as DC to EM; and hence, the limit of their ratio is equal to the limit of the ratio of DC to EM, which is equal to unity.

But the curvelinear area $DCME$ is less than the rectangle $DGME$, and greater than the rectangle $DCFE$: hence, the limit of its ratio to either of them will be unity. But,

$$\frac{DCME}{DE}=\frac{DCME}{DE}\times\frac{DEFC}{DEFC}=DC\times\frac{DCME}{DEFC},$$

or by representing the area of the segment by s and the ordinate DC by y, and passing to the limit, we have

$$\frac{ds}{dx}=y, \qquad \text{or} \qquad ds=ydx;$$

hence, *the differential of the area of a segment of any curve, is equal to the ordinate into the differential of the abscissa.*

131. To find the differential of the area of a circular segment, we have

$$x^2+y^2=R^2, \qquad \text{and} \qquad y=\sqrt{R^2-x^2};$$

hence, $$ds=dx\sqrt{R^2-x^2}.$$

The differential of the segment of an ellipse, is

$$ds=\frac{B}{A}dx\sqrt{A^2-x^2},$$

and of the segment of a parabola

$$ds=dx\sqrt{2px}.$$

Signification of the Differential Coefficients.

132. It has already been shown that, if the ordinate of a curve be regarded as a function of the abscissa, the first differential coefficient will be equal to the tangent of the angle which the tangent line forms with the axis of abscissas (Art. 113). We now propose to show the signification of the second differential coefficient, the ordinate being regarded as a function of the abscissa.

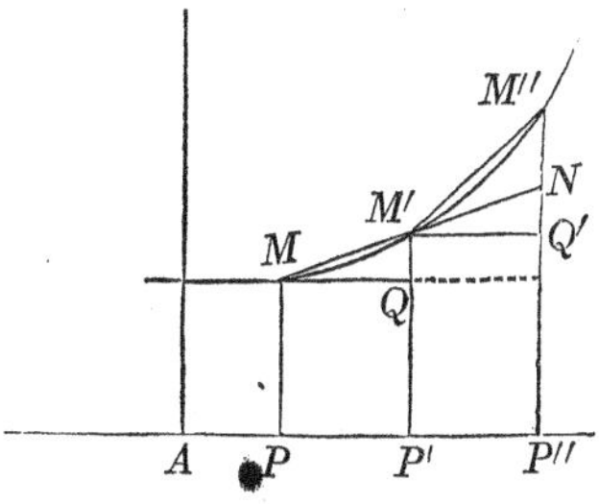

Let AP be the abscissa and PM the ordinate of a curve. From P lay off on the axis of abscissas $PP'=h$, and $PP''=2h$. Draw the ordinates PM, $P'M'$, $P''M''$; also the lines $MM'N$, $M'M''$; and lastly, MQ, $M'Q'$, parallel to the axis of abscissas. Then will $M'Q = NQ'$, and we shall have

$$PM = y,$$

$$P'M' = y + \frac{dy}{dx}\frac{h}{1} + \frac{d^2y}{dx^2}\frac{h^2}{1.2} + \&c.,$$

$$P''M'' = y + \frac{dy}{dx}\frac{2h}{1} + \frac{d^2y}{dx^2}\frac{4h^2}{1.2} + \&c.,$$

$$P'M' - PM = M'Q = \frac{dy}{dx}\frac{h}{1} + \frac{d^2y}{dx^2}\frac{h^2}{1.2} + \&c.,$$

$$P''M'' - P'M' = M''Q' = \frac{dy}{dx}h + \frac{d^2y}{dx^2}\frac{3h^2}{1.2} + \&c.$$

$$M''Q' - M'Q = +M''N = \frac{d^2y}{dx^2}h^2 + \&c.$$

Now, since the sign of the first member of the equation is essentially positive, the sign of the second member will also be positive (Alg. Art. 85). But by diminishing h, the sign of the second member will depend on that of the second differential coefficient (Art. 44): hence, the second differential coefficient is positive.

If the curve is below the axis of abscissas, the ordinates will be negative, and it is easily seen that we shall then have

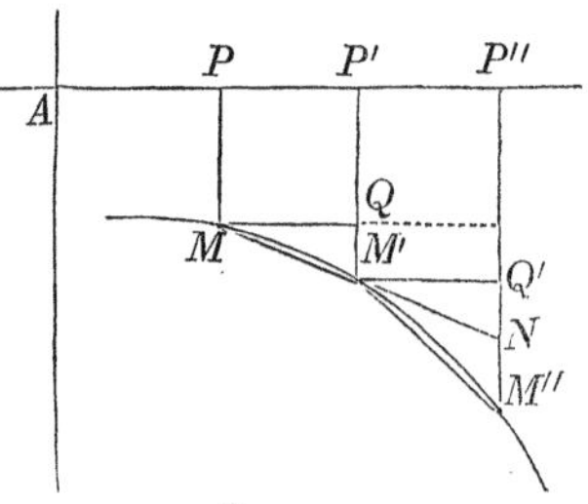

$$M''Q' - M'Q = -M''N = \frac{d^2y}{dx^2}h^2 + \&c.$$

Now, since the first member is negative, the second member will be negative: hence we conclude that, *if a curve is convex towards the axis of abscissas, the ordinate and second differential coefficient will have like signs.*

133. Let us now consider the curve $CMM'M''$, which is concave towards the axis of abscissas. We shall have,

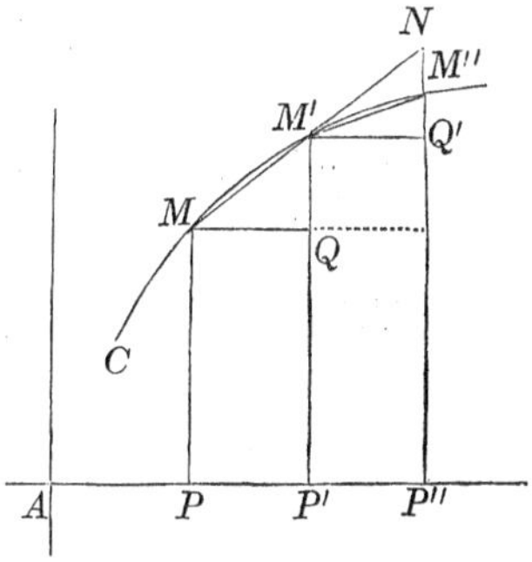

$$PM = y,$$

$$P'M' = y + \frac{dy}{dx}\frac{h}{1} + \frac{d^2y}{dx^2}\frac{h^2}{1.2} + \&c.,$$

$$P''M'' = y + \frac{dy}{dx}\frac{2h}{1} + \frac{d^2y}{dx^2}\frac{4h^2}{1.2} + \&c.,$$

$$P'M' - PM = M'Q = \frac{dy}{dx}\frac{h}{1} + \frac{d^2y}{dx^2}\frac{h^2}{1.2} + \&c.,$$

$$P''M'' - P'M' = M''Q' = \frac{dy}{dx}h + \frac{d^2y}{dx^2}\frac{3h^2}{1.2} + \&c.,$$

$$M''Q' - M'Q = -NM'' = \frac{d^2y}{dx^2}h^2 + \&c.$$

But since the first member of the equation is negative, the essential sign of the second member will also be negative: hence, the second differential coefficient will be negative.

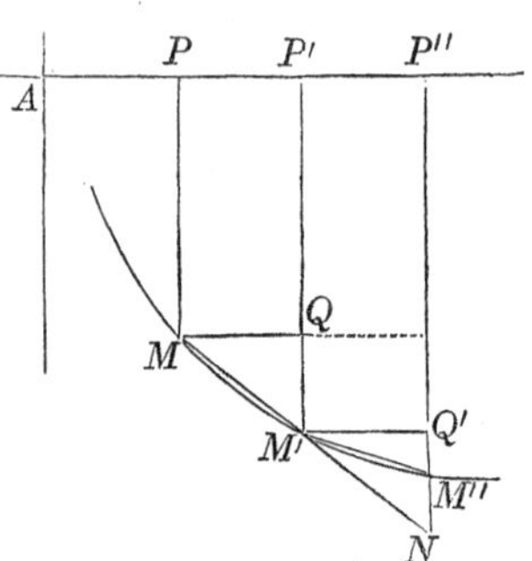

If the curve is below the axis of abscissas, the ordinate will be negative, and it is easily seen that we should then have

$$M''Q' - M'Q = +NM'' = \frac{d^2y}{dx^2}h^2 + \text{\&c.};$$

hence we conclude that, *if a curve is concave towards the axis of abscissas, the ordinate and second differential coefficient will have contrary signs.*

The ordinate will be considered as positive, unless the contrary is mentioned.

134. *Remark* 1. The co-ordinates x and y, determine a single point in a curve, as M. The differential of y is derived from the ordinate PM, and is what QM' becomes when the ordinates $P'M'$ and PM become consecutive.

The second differential of y is derived from $M'Q$, in the same way that dy is derived from the primitive function y. It is, indeed, what $M''Q'$ becomes, when $M''Q'$ becomes consecutive with $M'Q$. The abscissa x being supposed to increase uniformly, the difference between PP' and $P'P''$ is 0: and therefore the second differential of x is 0. The co-ordinates x and y, and the first and second differentials determine three points, M, M', M'', consecutive with each other.

135. *Remark* 2. When the curve is convex towards

the axis of abscissa, the first differential coefficient, which represents the tangent of the angle formed by the tangent line with the axis of abscissas, is an increasing function of the abscissa: hence, its differential coefficient, that is, the second differential coefficient of the function, ought to be positive (Art. 31).

When the curve is concave, the first differential coefficient is a decreasing function of the abscissa; hence, the second differential coefficient should be negative (Art. 31).

Examination of the Singular Points of Curves.

136. A singular point of a curve is one which is distinguished by some particular property not enjoyed by the points of the curve in general.

Let us, as a first example, find the points of a curve, through which the tangent lines will be parallel or perpendicular to the axis of abscissa.

137. Since the first differential coefficient expresses the value of the tangent of the angle which the tangent line forms with the axis of abscissas, and since the tangent is 0, when the angle is 0, and infinite when the angle is 90°, it follows that the roots of the equation

$$\frac{dy}{dx}=0,$$

will give the abscissas of all the points at which the tangent is parallel to the axis of abscissas, and the roots of the equation

$$\frac{dy}{dx}=\infty, \qquad \text{or} \qquad \frac{dx}{dy}=0,$$

will give the abscissas of all the points at which the tangent is perpendicular to the axis of abscissas.

138. If a curve from being convex towards the axis of abscissas becomes concave, or from being concave becomes convex, the point at which the change of curvature takes place is called a *point of inflexion.*

Since the ordinate and differential coefficient of the second order have the same sign when the curve is convex towards the axis of abscissas, and contrary signs when it is concave, it follows that at the point of inflexion, the second differential coefficient will change its sign. Therefore between the positive and negative values there will be one value of x which will reduce the second differential coefficient to 0 or infinity (Alg. Art. 310): hence the roots of the equations

$$\frac{d^2y}{dx^2}=0, \text{ or } \frac{d^2y}{dx^2}=\infty$$

will give the abscissas of the points of inflexion.

139. Let us now apply these principles in discussing the equation of the circle

$$x^2+y^2=R^2.$$

We have, by differentiating,

$$\frac{dy}{dx}=-\frac{x}{y},$$

and placing

$$-\frac{x}{y}=0, \qquad \text{we have} \qquad x=0.$$

Substituting this value in the equation of the curve, we have

$$y=\pm R;$$

hence, the tangent is parallel to the axis of abscissas at the two points where the axis of ordinates intersects the circumference.

If we make

$$\frac{dy}{dx} = -\frac{x}{y} = \infty, \qquad \text{or} \qquad -\frac{y}{x} = 0,$$

we have $y = 0$; substituting this value in the equation, we find

$$x = \pm R,$$

and hence, the tangent is perpendicular to the axis of abscissas at the points where the axis intersects the circumference.

The second differential coefficient is equal to

$$-\frac{R^2}{y^3},$$

which will be negative when y is positive, and positive when y is negative. Hence, the circumference of the circle is concave towards the axis of abscissas.

If we apply a similar analysis to the equation of the ellipse, we shall find the tangents parallel to the axis of abscissas at the extremities of one axis, and perpendicular to it at the extremities of the other, and the curve concave towards its axes.

140. Let us now discuss a class of curves, which may be represented by the equation

$$y = b \pm c(x - a)^m,$$

in which we suppose c to be positive or negative, and different values to be attributed to the exponent m.

1st. *When* c *is positive, and* m *entire and even.*

By differentiating, we have

$$\frac{dy}{dx} = mc(x-a)^{m-1},$$

$$\frac{d^2y}{dx^2} = m(m-1)c(x-a)^{m-2}.$$

If we place the value $\frac{dy}{dx} = 0$, we find $x = a$, and substituting this value in the equation of the curve, we find

$$y = b:$$

hence, $x = a$, $y = b$, are the co-ordinates of the point at which the tangent line is parallel to the axis of abscissas.

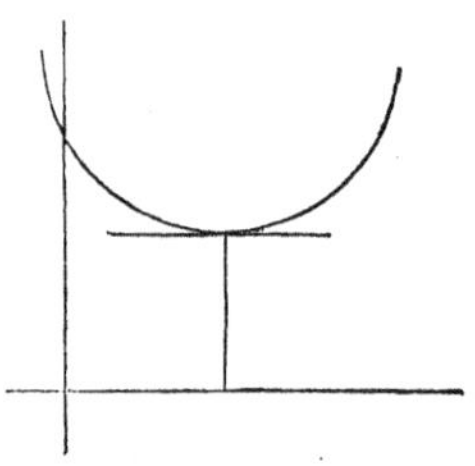

Since m is even, $m - 2$ will also be even, and hence the second differential coefficient will be positive for all values of x. The curve will therefore be convex towards the axis of X, and there will be no point of inflexion.

The value of $x = a$ renders the ordinate y a minimum, since after m differentiations a differential coefficient of an even order becomes constant and positive (Art. 110).

The curve does not intersect the axis of X, but cuts the axis of Y at a distance from the origin expressed by

$$y = b + ca^m.$$

141. 2d. *When* c *is negative, and* m *entire and even.*

We shall have, by differentiating, $y = b - c(x-a)^m$

$$\frac{dy}{dx} = -mc(x-a)^{m-1},$$

and $$\frac{d^2y}{dx^2} = -m(m-1)c(x-a)^{m-2}.$$

The discussion is the same as before, excepting that the second differential coefficient being negative for all values of x, the curve is concave towards the axis of abscissas, and the value of $x = a$, renders the ordinate y a maximum (Art. 110).

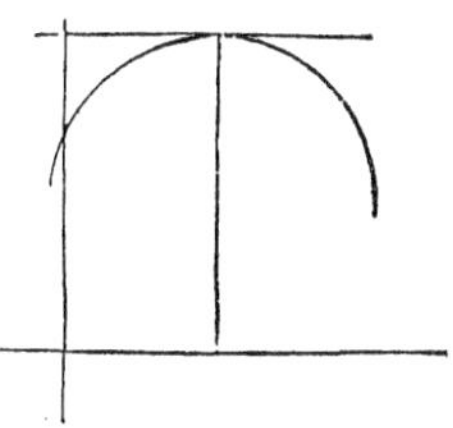

142. 3d. *When* c *is plus or minus, and* m *entire and uneven.*

We shall have, by differentiating,

$$\frac{dy}{dx} = \pm mc(x-a)^{m-1},$$

and $$\frac{d^2y}{dx^2} = \pm m(m-1)c(x-a)^{m-2}.$$

The first differential coefficient will be 0, when $x = a$, hence, the tangent will be parallel to the axis of abscissas, at the point of which the co-ordinates are $x = a$, $y = b$.

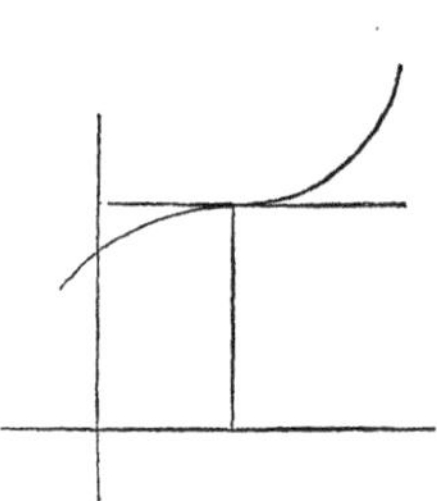

Since the exponent $m-2$ is uneven, the factor $(x-a)^{m-2}$ will be negative when $x<a$, and positive when $x>a$; hence, this factor changes its sign at the point of the curve of which the abscissa is $x=a$.

If c is positive, the second differential coefficient will be negative for $x<a$, and positive for $x>a$: hence there will be an inflexion when $x=a$. If c were negative, the curve would be first convex and then concave towards the axis of abscissas, but there would still be an inflexion at the point $x=a$. At this point the tangent line separates the two branches of the curve.

There will, in this case, be neither a maximum nor a minimum, since after m differentiations a differential coefficient of an odd order, will become equal to a constant quantity (Art. 110).

143. 4th. *When* c *is positive or negative, and* m *a fraction having an even numerator, as* $m=\frac{2}{3}$.

By differentiating, and supposing c positive, we have

$$\frac{dy}{dx}=\frac{2}{3}c(x-a)^{\frac{2}{3}-1}=\frac{2c}{3(x-a)^{\frac{1}{3}}},$$

$$\frac{d^2y}{dx^2}=-\frac{2c}{9(x-a)^{\frac{4}{3}}},$$

If we make $x=a$, the first differential coefficient will become infinite; and the tangent will be perpendicular to the axis of abscissas, at the point of which the co-ordinates are $x=a$, $y=b$.

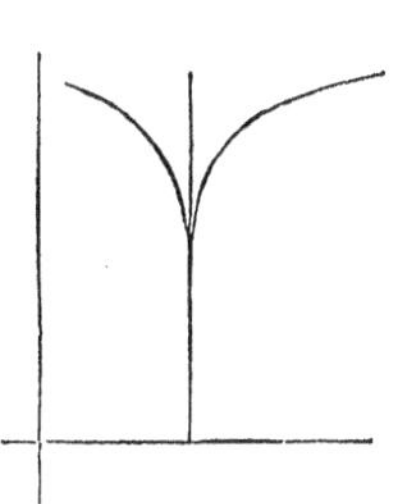

In regard to the second differential coefficient, it will become infinite for $x=a$, and negative for every other value of x, since the factor $(x-a)$ of the denominator is raised to a power denoted by an even exponent. Hence, the curve will be concave towards the axis of abscissas.

If we take the equation of the curve

$$y=b+c(x-a)^{\frac{2}{3}},$$

and make $x=a+h$, and $x=a-h$, we shall have, in either case,

$$y=b+ch^{\frac{2}{3}};$$

and hence, y will be less for $x=a$, than for any other value of x, either greater or less than a. Hence, the value $x=a$, renders y a *minimum*.

If c were negative, the equation would be of the form

$$y=b-c(x-a)^{\frac{2}{3}};$$

and we should have, by differentiating,

$$\frac{dy}{dx}=-\frac{2c}{3(x-a)^{\frac{1}{3}}},$$

and

$$\frac{d^2y}{dx^2}=\frac{2c}{9(x-a)^{\frac{4}{3}}}.$$

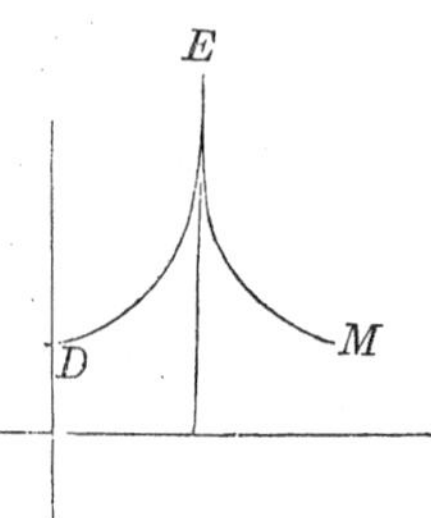

The first and second differential coefficients will be infinite for $x = a$, and the second differential coefficient will be positive for all values of x greater or less than a; and hence, the curve will be convex towards the axis of abscissas.

If, in the equation of the curve

$$y = b - c(x-a)^{\frac{2}{3}},$$

we make $x = a + h$, and $x = a - h$, we shall have, in either case,

$$y = b - ch^{\frac{2}{3}};$$

and hence, y will be greater for $x = a$, than for any other value of x either greater or less than a. Hence, the value $x = a$, renders y a maximum.

144. *Remark.* The conditions of a maximum or a minimum deduced in Art. 110, were established by means of the theorem of Taylor. Now, the case in which the function changes its form by a particular value attributed to x, was excluded in the demonstration of that theorem (Art. 45). Hence, the conditions of minimum and maximum deduced in the two last cases, ought not to have appeared among the general conditions of Art. 110.

We therefore see that there are two species of maxima and minima, the one characterized by

$$\frac{dy}{dx} = 0, \quad \text{the other by} \quad \frac{dy}{dx} = \infty.$$

In the first, we determine whether the function is a maximum or a minimum by examining the subsequent differential coefficient; and in the second, by examining the value of the function before and after that value of x which renders the first differential coefficient infinite.

The branches DE, ME, which are both represented by the equation.

$$y = b \pm c(x - a)^{\frac{2}{3}},$$

are not considered as parts of a continuous curve. For, the general relations between y and x which determine each of the parts DE, ME, is entirely broken at the point M, where $x = a$. The two parts are therefore regarded as separate branches which unite at M. The point of union is called a *cusp*, or a *cusp point*.

145. 5th. *When* c *is positive or negative and* m *a fraction having an even denominator, as* m $= \frac{3}{4}$.

Under this supposition the equation of the curve will become

$$y = b \pm c(x - a)^{\frac{3}{4}},$$

and by differentiating, we have

$$\frac{dy}{dx} = \pm \frac{3c}{4(x - a)^{\frac{1}{4}}},$$

and

$$\frac{d^2y}{dx^2} = \mp \frac{3c}{4 . 4(x - a)^{\frac{5}{4}}}.$$

The curve represented by this eouation will have two branches: the one corresponding to the plus sign will be concave towards the axis of abscissas, and the one corresponding to the minus sign will be convex. Every value of x less than a will render y imaginary. The co-ordinates of the point M, are $x = a$, $y = b$.

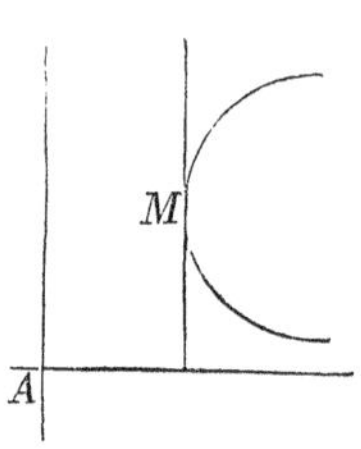

146. 6th. *When* c *is positive or negative and* m *a fraction having an uneven numerator and an uneven denominator, as* $m = \frac{3}{5}$.

Under this supposition the equation will become

$$y = b \pm c(x - a)^{\frac{3}{5}},$$

and by differentiating, we have

$$\frac{dy}{dx} = \pm \frac{3c}{5(x - a)^{\frac{2}{5}}},$$

$$\frac{d^2y}{dx^2} = \mp \frac{3.2c}{5.5(x - a)^{\frac{7}{5}}};$$

from which we see that if we use the superior sign of the first equation, the curve will be convex towards the axis of abscissas for $x < a$, that there will be a point of inflexion for $x = a$, and that the curve will be concave for $x > a$. If the lower sign be employed, the first branch will become concave, and the other convex.

147. The cusps, which have been considered, were formed by the union of two curves that were convex to-

wards each other, and such are called, *cusps of the first order.*

It frequently happens, however, that the curves which unite, embrace each other. The equation

$$(y - x^2)^2 = x^5,$$

furnishes an example of this kind. By extracting the square root of both members and transposing, we have

$$y = x^2 \pm x^{\frac{5}{2}};$$

and by differentiating

$$\frac{dy}{dx} = 2x \pm \frac{5}{2}x^{\frac{3}{2}}, \qquad \frac{d^2y}{dx^2} = 2 \pm \frac{5}{2}.\frac{3}{2}x^{\frac{1}{2}}.$$

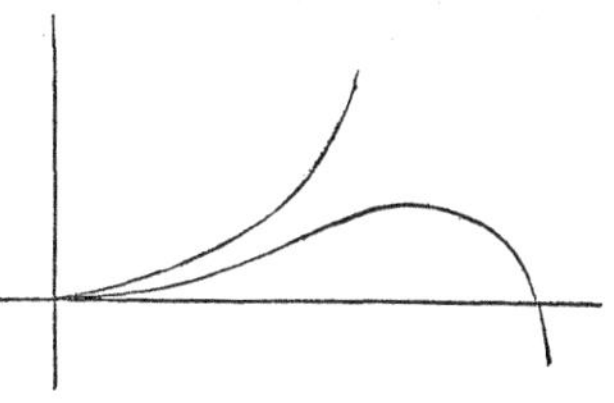

We see by examining the equations, that the curve has two branches, both of which pass through the origin of co-ordinates. The upper branch, which corresponds to the plus sign, is constantly convex towards the axis of abscissas, while the lower branch is convex for $x < \frac{64}{225}$, and concave for $x > \frac{64}{225}$ and $x < 1$. At the last point the curve passes below the axis of abscissas and becomes convex towards it. If we make the first differential coefficient equal to 0, we shall find $x = 0$, and substituting this value in the equation of the curve, gives $y = 0$; and hence, the axis of abscissas is tangent to both branches of the curve at the origin of co-ordinates. At this point the differential coefficient of the second order is positive for both branches of the curve, hence they are both convex towards the axis. When the cusp is formed by the union of two curves which, at the point of contact, lie on the same side of the common tangent, it is called a cusp of the *second order*.

148. Let us, as another example, discuss the curve whose equation is

$$y = b \pm (x - a)\sqrt{x - c}.$$

By differentiating, we obtain

$$\frac{dy}{dx} = \pm\sqrt{x-c} \pm \frac{x-a}{2\sqrt{x-c}}.$$

We see, from the equation of the curve, that y will be imaginary for all values of x less than c.

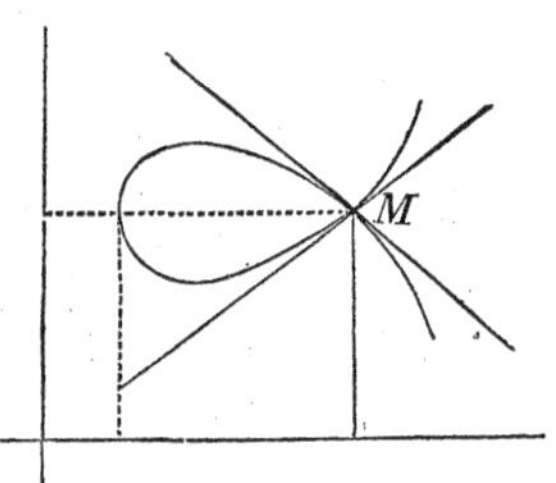

For $x = c$, we have $y = b$; and for $x > c$, we have two values of y and consequently two branches of the curve, until $x = a$ when they unite at the point M. For $x > a$ there will be two real values of y and consequently two branches of the curve. The point M, at which the branches intersect each other, is called a *multiple point*, and differs from a cusp by being a point of intersection instead of a point of tangency. At the multiple point M there are two tangents, one to each branch of the curve. The one makes an angle with the axis of abscissas, whose tangent is

$$+\sqrt{a-c} \qquad ;$$

the other, an angle whose tangent is

$$-\sqrt{a-c}\cdot$$

149. Besides the cusps and multiple points which have already been discussed, there are sometimes other points lying entirely without the curve, and having no connexion with it, excepting that their co-ordinates will satisfy the equation of the curve.

For example, the equation

$$ay^2 - x^3 + bx^2 = 0,$$

will be satisfied for the values $x = \pm 0$, $y = \pm 0$; and hence, the origin of co-ordinates A, satisfies the equation of the curve, and enjoys the property of a multiple point, since it is the point of union of two values of x, and two values of y.

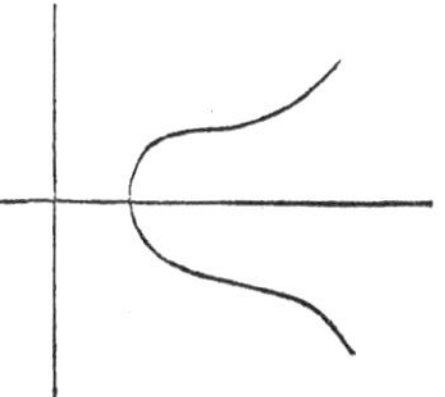

If we resolve the equation with respect to y, we find

$$y = \pm x\sqrt{\frac{x-b}{a}};$$

and hence, y will be imaginary for all negative values of x, and for all positive values between the limits $x = 0$ and $x = b$. For all positive values of x greater than b, the values of y will be real.

The first differential coefficient is

$$\frac{dy}{dx} = \frac{x(3x-2b)}{2\sqrt{ax^2(x-b)}};$$

or by dividing by the common factor x,

$$\frac{dy}{dx}=\frac{3x-2b}{2\sqrt{a(x-b)}}$$

and making $x=0$, there results

$$\frac{dy}{dx}=-\frac{2b}{2\sqrt{-ab}},$$

which is imaginary, as it should be, since there is no poir. of the curve which is consecutive with the isolated or conjugate point. The differential coefficients of the higher orders are also imaginary at the conjugate points.

150. We may draw the following conclusions from the preceding discussion.

1st. The equation $\frac{dy}{dx}=0$, determines the points at which the tangents are parallel to the axis of abscissas.

2d. The equation $\frac{dy}{dx}=\infty$, determines the points of the curve at which the tangents are perpendicular to the axis of abscissas. The two last equations also determine the cusps, if there are any, in all cases where the tangent at the cusps is parallel or perpendicular to the axis of abscissas.

3d. The equation $\frac{d^2y}{dx^2}=0$, or $\frac{d^2y}{dx^2}=\infty$ determines the points of inflexion.

4th. The equation $\frac{dy}{dx}=$ an imaginary constant, indicates a conjugate point.

CHAPTER VII.

Of Osculatory Curves—Of Evolutes.

151. Let PT be tangent to the curve ABP at the point P, and PN a normal at the same point: then will PT be tangent to the circumference of every circle passing through P, and having its centre in the normal PN.

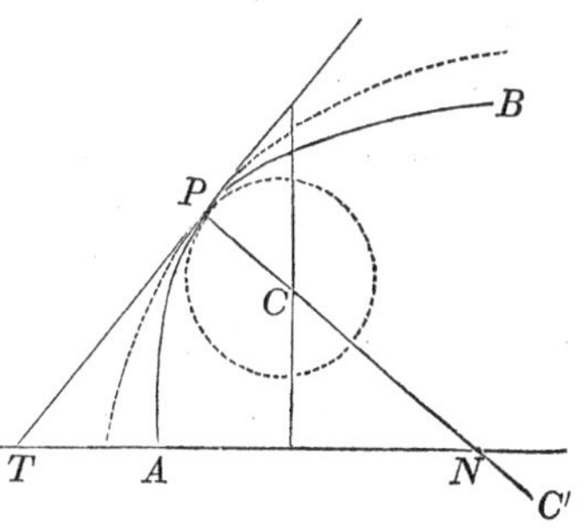

It is plain that the centre of a circle may be taken at some point C, so near to P, that the circumference shall fall within the curve APB, and then every circumference described with a less radius, will fall entirely within the curve. It is also apparent, that the centre may be taken at some point C', so remote from P, that the circumference shall fall between the curve APB and the tangent PT, and then every circumference described with a greater radius will fall without the curve. Hence, there are two classes of tangent circles which may be described; the one lying within the curve, and the other without it.

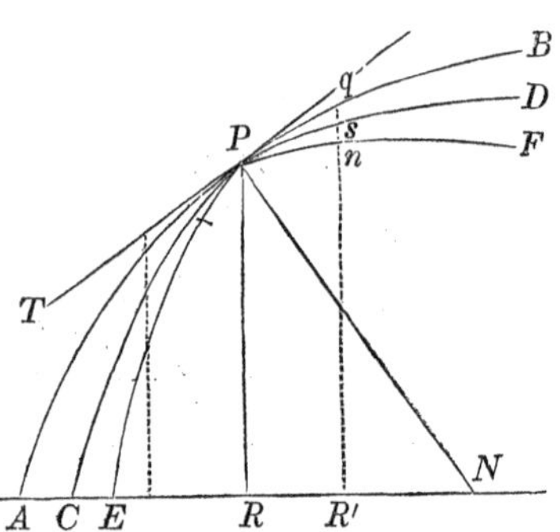

152. Let there be three curves, APB, CPD, EPF, which have a common tangent TP, and a common normal PN; then will they be tangent to each other at the point P. It does not follow, however, from this circumstance, that each curve will have an *equal tendency* to coincide with the tangent TP, nor does it follow that any two of the curves CPD, EPF, will have an equal tendency to coincide with the first curve APB.

It is now proposed to establish the analytical conditions which determine the tendency of curves to coincide with each other, or with a common tangent.

Designate the co-ordinates of the first curve APB by x and y, the co-ordinates of the second CPD by x', y', and the co-ordinates of the third EPF by x'', y''. If we designate the common ordinate PR by y, y', y'', we shall then have

$$qR' = y + \frac{dy}{dx}\frac{h}{1} + \frac{d^2y}{dx^2}\frac{h^2}{1.2} + \frac{d^3y}{dx^3}\frac{h^3}{1.2.3} + \&c.,$$

$$sR' = y' + \frac{dy'}{dx'}\frac{h}{1} + \frac{d^2y'}{dx'^2}\frac{h^2}{1.2} + \frac{d^3y'}{dx'^3}\frac{h^3}{1.2.3} + \&c.;$$

$$nR' = y'' + \frac{dy''}{dx''}\frac{h}{1} + \frac{d^2y''}{dx''^2}\frac{h^2}{1.2} + \frac{d^3y''}{dx''^3}\frac{h^3}{1.2.3} + \&c.$$

But since the curves are tangent to each other at the point P, we have (Art. 119),

$$y = y' = y'', \quad \text{and} \quad \frac{dy}{dx} = \frac{dy'}{dx'} = \frac{dy''}{dx''}: \quad \text{hence,}$$

$$d = qs = \left(\frac{d^2y}{dx^2} - \frac{d^2y'}{dx'^2}\right)\frac{h^2}{1.2} + \left(\frac{d^3y}{dx^3} - \frac{d^3y'}{dx'^3}\right)\frac{h^3}{1.2.3} + \text{\&c.},$$

$$d' = qn = \left(\frac{d^2y}{dx^2} - \frac{d^2y''}{dx''^2}\right)\frac{h^2}{1.2} + \left(\frac{d^3y}{dx^3} - \frac{d^3y''}{dx''^2}\right)\frac{h^3}{1.2.3} + \text{\&c.}$$

Now, in order that the first curve APB shall approach more nearly to the second CPD than to the third EPF, we must have

$$d < d',$$

and consequently,

$$A\frac{h^2}{1.2} + B\frac{h^3}{1.2.3} + \text{\&c.}, < A'\frac{h^2}{1.3} + B'\frac{h^3}{1.2.3} + \text{\&c.},$$

in which we have represented the coefficients in the first series by A, B, C, &c., and the coefficients in the second by A', B', C', &c.

Now, the limit of the first member of the inequality will always be less than the limit of the second, when its first term involves a higher power of h than the first term of the second. For, if $A = 0$, the first member will involve the highest power of h, and we shall have

$$B\frac{h^3}{1.2.3} + \text{\&c.}, < A'\frac{h^2}{1.2} + B'\frac{h^3}{1.2.3} + \text{\&c.},$$

and by dividing by h^2,

$$B\frac{h}{1.2.3} + \text{\&c.}, < A'\frac{1}{1.2} + B'\frac{h}{1.2.3} + \text{\&c.},$$

and by passing to the limit

$$0 < A' \frac{1}{1.2}.$$

But when $A = 0$, we have

$$\frac{d^2y}{dx^2} = \frac{d^2y'}{dx'^2},$$

and hence, when three curves have a common ordinate, the first will approach nearer to the second than to the third, *if the number of equal differential coefficients between the first and second is greater than that between the first and third.* And consequently, if the first and second curves have $m+1$ differential coefficients which are equal to each other, and the first and third curves only m equal differential coefficients, the first curve will approach more nearly to the second than to the third. Hence it appears, that the order of contact of two curves will depend on the number of corresponding differential coefficients which are equal to each other.

The contact which results from an equality between the co-ordinates and the first differential coefficients, is called a contact of the *first order*, or a simple tangency (Art. 119). If the second differential coefficients are also equal to each other, it is called a contact of the *second order.* If the first three differential coefficients are respectively equal to each other, it is a contact of the *third order;* and if there are m differential coefficients respectively equal to each other, it is a contact of the m*th order.*

153. Let us now suppose that the second line is only given in species, and that values may be attributed at pleasure to the constants which enter its equation. We

shall then be able to establish between the first and second lines as many conditions as there are constants in the equation of the second line. If, for example, the equation of the second line contains two constants, two conditions can be established, viz.: an equality between the co-ordinates, and an equality between the first differential coefficients; this will give a contact of the first order.

If the equation of the second curve contains three constants, three conditions may be established, viz.: an equality between the co-ordinates, and an equality between the first and second differential coefficients. This will give a contact of the second order. If there are four constants, we can obtain a contact of the third order; and if there are $m+1$ constants, a contact of the mth order.

It is plain, that in each of the foregoing cases the highest order of contact is determined.

The line which has a higher order of contact with a given curve than can be found for any other line of the same species, is called an osculatrix.

Let it be required, for example, to find a straight line which shall be osculatory to a curve, at a given point of which the co-ordinates are x'', y''.

The equation of the right line is of the form

$$y = ax + b,$$

and it is required to find such values for the constants a and b as to cause the line to fulfil the conditions,

$$x = x'', \qquad y = y'', \qquad \text{and} \qquad \frac{dy}{dx} = \frac{dy''}{dx''}.$$

By differentiating the equation of the line, we have

$$\frac{dy}{dx} = a;$$

and since the line passes through the point of osculation

$$y - y'' = \frac{dy}{dx}(x - x'').$$

Substituting for $\frac{dy}{dx}$ its value $\frac{dy''}{dx''}$, we have

$$y - y'' = \frac{dy''}{dx''}(x - x''),$$

for the equation of the osculatrix.

In the equation of the circle

$$x^2 + y^2 = R^2,$$

we find $\quad \frac{dy}{dx} = -\frac{x}{y} = \frac{dy''}{dx''} = -\frac{x''}{y''}$

hence, the equation of the osculatrix of the first order, to the circle, is

$$y - y'' = -\frac{x''}{y''}(x - x''),$$

or by reducing $\quad yy'' + xx'' = R^2.$

154. If α and β represent the co-ordinates of the centre of a circle, its equation will be of the form

$$(x - \alpha)^2 + (y - \beta)^2 = R^2.$$

If this equation be twice differentiated, we shall have,

$$(x - \alpha)dx + (y - \beta)dy = 0,$$

$$dx^2 + dy^2 + (y - \beta)d^2y = 0;$$

and by combining the three equations, we obtain,

$$y - \beta = -\frac{dx^2 + dy^2}{d^2y},$$

$$x - \alpha = \frac{dy}{dx}\left(\frac{dx^2 + dy^2}{d^2y}\right),$$

$$R = \pm\frac{(dx^2 + dy^2)^{\frac{3}{2}}}{dx\, d^2y}.$$

If it be now required to make this circle osculatory to a given curve, at a point of which the co-ordinates are x'', y'', we have only to substitute in the three last equations, the values of

$$\frac{dy}{dx} = \frac{dy''}{dx''}, \qquad \frac{d^2y}{dx^2} = \frac{d^2y''}{dx''^2},$$

deduced from the equation of the curve, and to suppose, at the same time, the co-ordinates x and y in the curve to become equal to those of x and y in the circle.

If we suppose x'', y'' to become general co-ordinates of the curve, the circle will move around the curve, continually changing its radius, and will become osculatory at all the points in succession.

155. If the circle CD be osculatory to the curve EF, at the point P, we shall have

$$qs = C \times + \frac{h^3}{1.2.3} + \&c.,$$

for h positive; and

$$q's' = C \times - \frac{h^3}{1.2.3} + \&c.,$$

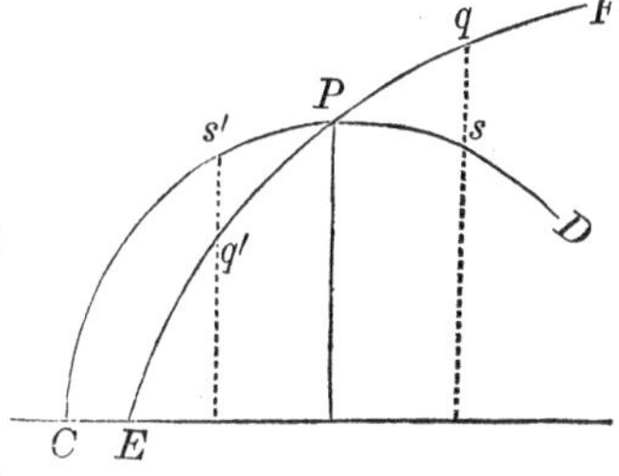

for h negative: hence, the two lines qs, $q's'$, have contrary signs. The curve, therefore, lies above the osculatory circle on one side of the point P, and below it on the other, and consequently, divides the osculatory circle at the point of osculation. Hence, also, the osculatory circle separates the tangent circles which lie without the curve from those which lie within it (Art. 151).

In every osculatrix of an even order the first term in the values of qs, $q's'$, will, in general, contain an uneven power of h; and hence their signs may be made to depend on that of h. The curve will therefore lie above the osculatrix on one side of the point P, and below it on the other; and hence, *every osculatrix of an even order, will in general be divided by the curve at the point of osculation.*

156. The first differential equation of Article 154,

$$(x-\alpha)dx+(y-\beta)dy=0$$

may be placed under the form

$$\beta-y=-\frac{dx}{dy}(\alpha-x).$$

If we make the circle osculatory to the curve we have

$$x=x'', \quad y=y'', \quad \text{and}$$

$$\frac{dx}{dy}=\frac{dx''}{dy''}; \text{ hence,}$$

$$\beta-y''=-\frac{dx''}{dy''}(\alpha-x''),$$

which is the equation of a normal at the point whose co-ordinates are x'' y'' (Art. 122). But this normal passes through the point whose co-ordinates are α and β. Hence, *the normal drawn through the point of osculation, will contain the centre of the osculatory circle.*

157. It was shown in (Art. 155) that the osculatory circle is, in general, divided by the curve at the point of oscu-

lation. The position of the curves with respect to each other indicates this result.

For, the osculatory circle is always symmetrical with respect to the normal, while the curve is, in general, not symmetrical with respect to this line. If, however, the curve is symmetrical with respect to the normal, as is the case in lines of the second order when the normal coincides with an axis, the curve will not divide the osculatory circle at the point of osculation; and the condition which renders the second differential coefficients in the curve and circle equal to each other, will also render the third differential coefficients equal, and the contact will then be of the third order.

158. The radius of the osculatory circle

$$R = \pm \frac{(dx^2 + dy^2)^{\frac{3}{2}}}{dxd^2y}$$

is affected with the sign plus or minus, and it may be well to determine the circumstances under which each sign is to be used.

If we suppose the ordinate to be positive, we shall have (Art. 133)

$$\frac{d^2y}{dx^2}, \text{ and consequently } d^2y$$

negative when the curve is concave towards the axis of abscissas, and positive when it is convex. If then, we wish the radius of the osculatory circle to be positive for curves which are concave towards the axis of abscissas, we must employ the minus sign, in which case the radius will be negative for curves which are convex.

159. If the circumferences of two circles be described with different radii, and a tangent line be drawn to each, it is plain that the circumference which has the less radius will depart more rapidly from its tangent than the circumference which is described with the greater radius; and hence we say, that its *curvature is greater*. And generally, the curvature of any curve is said to be greater or less than that of another curve, according as its tendency to depart from its tangent at a given point, is greater or less than that of the curve with which it is compared.

160. The curvature is the same at all the points of the same circumference, and also in all circumferences described with equal radii, since the tendency to depart from the tangent is the same. In different circumferences, the curvature is measured by the angle formed by two radii drawn through the extremities of an arc of a given length.

Let r and r' designate the radii of two circles, a the length of a given arc measured on the circumference of each; c the angle formed by the two radii drawn through the extremities of the arc in the first circle, and c' the angle formed by the corresponding radii of the second. We shall then have

$$2\pi r \ : \ a \ :: \ 360^\circ \ : \ c, \quad \text{hence,} \quad c = \frac{360^\circ a}{2\pi r};$$

also,

$$2\pi r' \ : \ a \ :: \ 360^\circ \ : \ c', \quad \text{hence,} \quad c' = \frac{360^\circ a}{2\pi r'};$$

and consequently

$$c \ : \ c' \ :: \ \frac{1}{r} \ : \ \frac{1}{r'},$$

that is, *the curvature in different circumferences varies inversely as the radii.*

161. The curvature of plane curves is measured by means of the osculatory circle.

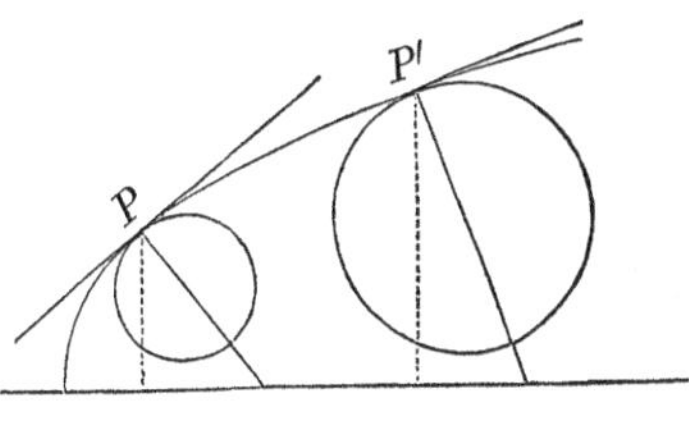

If we assume two points P and P', either on the same or on different curves, and find the radii r and r' of the circles which are osculatory at these points, then

$$\text{curvature at } P \ : \ \text{curvature at } P' \ :: \ \frac{1}{r} \ : \ \frac{1}{r'};$$

that is, *the curvature at different points varies inversely as the radius of the osculatory circle.*

The radius of the osculatory circle is called *the radius of curvature.*

162. Let us now determine the value of the radius of curvature for lines of the second order.

The general equation of these lines (An. Geom. Bk. VI, Prop. XII, Sch. 3), is

$$y^2 = mx + nx^2,$$

which gives,

$$dy = \frac{(m + 2nx)dx}{2y}, \quad dx^2 + dy^2 = \frac{[4y^2 + (m + 2nx)^2]\,dx^2}{4y^2}.$$

$$d^2y = \frac{2ny\,dx^2 - (m + 2nx)\,dx\,dy}{2y^2} = \frac{[4ny^2 - (m + 2nx)^2]\,dx^2}{4y^3}.$$

Substituting these values in the equation

$$R = -\frac{(dx^2 + dy^2)^{\frac{3}{2}}}{dx\,d^2y},$$

we obtain

$$R = \frac{[4(mx + nx^2) + (m + 2nx)^2]^{\frac{3}{2}}}{2m^2},$$

which is the general value of the radius of curvature in lines of the second order, for any abscissa x.

163. If we make $x = 0$, we have

$$R = \frac{1}{2}m = \frac{B^2}{A};$$

that is, in lines of the second order, *the radius of curvature at the vertex of the transverse axis is equal to half the parameter of that axis.*

If be required to find the value of the radius of curvature at the extremity of the conjugate axis of an ellipse, we make (An. Geom. Bk. VIII, Prop. XXI, Sch. 3),

$$m = \frac{2B^2}{A}, \qquad n = -\frac{B^2}{A^2}, \qquad \text{and} \qquad x = A,$$

which gives, after reducing,

$$R = \frac{A^2}{B}:$$

hence, *the radius of curvature at the vertex of the conjugate axis of an ellipse is equal to half the parameter of that axis.*

In the case of the parabola, in which $n = 0$, the general value of the radius of curvature becomes

$$R = \frac{(m^2 + 4mx)^{\frac{3}{2}}}{2m^2}.$$

164. If we compare the value of the radius of curvature with that of the normal line found in (Art. 118), we shall have

$$R = \frac{(\text{normal})^3}{\frac{1}{4}m^2};$$

that is, *the radius of curvature at any point is equal to the cube of the normal divided by half the parameter squared:* and hence, *the radii of curvature at different points of the same curve are to each other as the cubes of the corresponding normals.*

Of the Evolutes of Curves.

165. If we suppose an osculatory circle to be drawn at each of the points of the curve $APP'B$, and then a curve $ACC'C''$ to be drawn through the centres of these circles, this latter curve is called the *evolute curve*, and the curve $APP'B$ the *involute.*

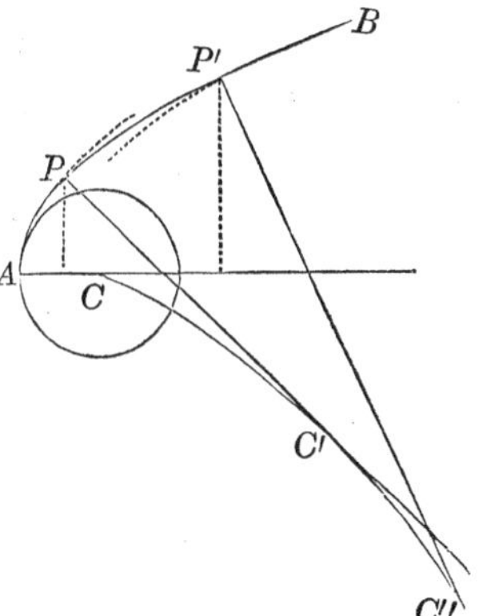

166. The co-ordinates of the centre of the osculatory circle, which have been represented by α and β, are constant for given values of the co-ordinates x and y of the

involute curve, but they become variable when we pass from one point of the involute curve to another.

167. We have already seen that the osculatory circle is characterized by the equations (Art. 154)

$$(x-\alpha)^2+(y-\beta)^2=R^2, \quad (1)$$

$$(x-\alpha)\,dx+(y-\beta)\,dy=0, \quad (2)$$

$$dx^2+dy^2+(y-\beta)\,d^2y=0. \quad (3)$$

If it be required to find the relations between the co-ordinates of the involute and the co-ordinates of the evolute curves, we must differentiate equations (1) and (2) under the supposition that α and β, as well as x and y, are variables. We shall then have

$$(x-\alpha)\,dx+(y-\beta)\,dy-(x-\alpha)\,d\alpha-(y-\beta)\,d\beta=R\,dR,$$

$$dx^2+dy^2+(y-\beta)\,d^2y-d\alpha\,dx-d\beta\,dy=0.$$

Combining these with equations (2) and (3), we obtain

$$-(y-\beta)\,d\beta-(x-\alpha)\,d\alpha=R\,dR, \quad (4)$$

$$-d\alpha\,dx-d\beta\,dy=0.$$

The last equation gives

$$\frac{d\beta}{d\alpha}=-\frac{dx}{dy}. \quad (5)$$

But equation (2) may be placed under the form

$$y-\beta=-\frac{dx}{dy}(x-\alpha),$$

which represents a normal to the involute (Art. 122), and which becomes, by substituting for $-\dfrac{dx}{dy}$ its value $\dfrac{d\beta}{d\alpha}$,

$$y-\beta=\frac{d\beta}{d\alpha}(x-\alpha), \quad (6)$$

or
$$\beta-y=\frac{d\beta}{d\alpha}(\alpha-x) \qquad \text{(Art. 120).}$$

This last equation, which is but another form for the equation of the normal to the involute, is, in fact, the equation of a tangent line to the evolute, at the point whose co-ordinates are α and β; hence, *a normal line to the involute curve is tangent to the evolute.*

168. It is now proposed to show, that the radius of curvature and the evolute curve have equal differentials

Combining equations (2) and (5) we obtain

$$(x-\alpha)=(y-\beta)\frac{d\alpha}{d\beta}, \quad (7)$$

or by squaring both members,

$$(x-\alpha)^2=(y-\beta)^2\frac{d\alpha^2}{d\beta^2};$$

combining this last with equation (1) we have

$$\frac{(d\alpha^2+d\beta^2)}{d\beta^2}(y-\beta)^2=R^2. \quad (8)$$

Combining equations (4) and (7), we have

$$-(y-\beta)\,d\beta-(y-\beta)\frac{d\alpha^2}{d\beta}=R\,dR,$$

or
$$-\frac{(d\alpha^2+d\beta^2)}{d\beta}(y-\beta)=R\,dR;$$

or by squaring both members

$$\frac{(d\alpha^2 + d\beta^2)^2}{d\beta^2}(y - \beta)^2 = R^2(dR)^2.$$

Dividing this last by equation (8), member by member we have

$$(dR)^2 = d\alpha^2 + d\beta^2$$

or
$$dR = \sqrt{d\alpha^2 + d\beta^2}.$$

But if s represents the arc of the evolute curve, of which the co-ordinates are α and β, we shall have (Art. 128),

$$ds = \sqrt{d\alpha^2 + d\beta^2};$$

hence,
$$dR = ds;$$

that is, *the differential of the radius of curvature is equal to the differential of the arc of the evolute.*

169. It does not follow, however, from the last equation, that the radius of curvature is equal to the arc of the evolute curve, but only that one of them is equal to the other plus or minus a constant (Art. 22). Hence,

$$R = s + a$$

is the form of the equation which expresses the relation between them.

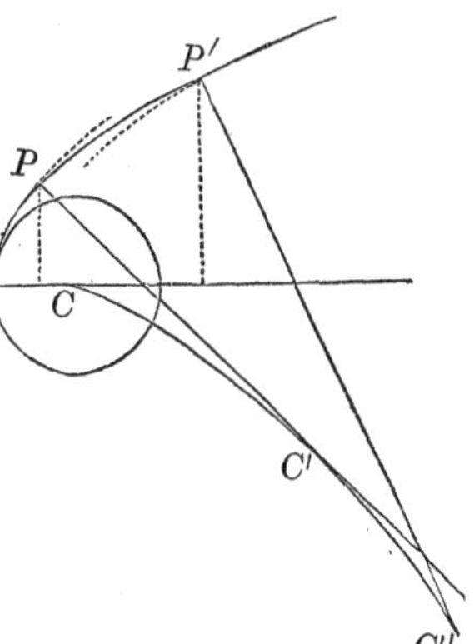

If we determine the radii of curvature at two points of the involute, as P and P', we shall have, for the first,

$$R = s + a,$$

and for the second

$$R' = s' + a;$$

hence,

$$R' - R = s' - s = C'C'';$$

and hence, *the difference between the radii of curvature at any two points of the involute is equal to the part of the evolute curve intercepted between them.*

170. The value of the constant a will depend on the position of the point from which the arc of the evolute curve is estimated.

If, for example, we take the radius of curvature for lines of the second order, and estimate the arc of the evolute curve from the point at which it meets the axis, the value of s will be 0 when $R = \frac{1}{2}m$ (Art. 163): hence we shall have

$$\frac{1}{2}m = 0 + a \quad \text{or} \quad a = \frac{1}{2}m;$$

and for any other point of the curve

$$R = s + \frac{1}{2}m.$$

Either of the evolutes, FE, FE', $F'E'$, or $F'E$, corresponding to one quarter of the ellipse, is equal to (Art. 169)

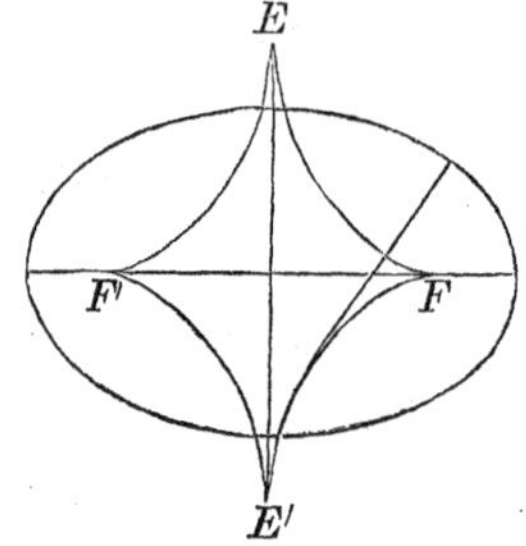

$$\frac{A^2}{B} - \frac{B^2}{A}.$$

171. The evolute curve takes its name from the connexion which it has with the corresponding involute.

Let $CC'C''$ be an evolute curve. At C draw a tangent AC, and make it equal to the constant a in the equation

$$R = s + a.$$

Wrap a thread $ACC'C''$ around the curve, and fasten it at any point, as C''.

Then, if we begin at A, and unwrap or *evolve* the thread, it will take the positions PC', $P'C''$, &c., and the point A will describe the involute APP': for

$$PC' - AC = CC' \quad \text{and} \quad P'C'' - AC = CC'C'', \text{ \&c.} \ldots$$

172. The equation of the evolute may be readily found by combining the equations

$$y - \beta = -\frac{dx^2 + dy^2}{d^2y}, \qquad x - \alpha = \frac{dy\,(dx^2 + dy^2)}{dx d^2y},$$

with the equation of the involute curve.

1st. Find, from the equation of the involute, the values of

$$\frac{dy}{dx} \quad \text{and} \quad d^2y,$$

and substitute them in the two last equations, and there will be obtained two new equations involving α, β, x and y.

2d. Combine these equations with the equation of the involute, and eliminate x and y: the resulting equation will contain α, β, and constants, and will be the equation of the evolute curve.

173. Let us take, as an example, the common parabola of which the equation is

$$y^2 = mx.$$

We shall then have

$$\frac{dy}{dx} = \frac{m}{2y}, \qquad d^2y = -\frac{m^2dx^2}{4y^3},$$

and hence

$$y - \beta = \frac{4y^3}{m^2}\left(\frac{4y^2 + m^2}{4y^2}\right) = \frac{4y^3 + m^2y}{m^2} = \frac{4y^3}{m^2} + y;$$

and by observing that the value of $x - \alpha$ is equal to that of $y - \beta$ multiplied by $-\frac{dy}{dx}$, we have

$$x - \alpha = -\frac{4y^2 + m^2}{2m};$$

hence we have,

$$-\beta = \frac{4y^3}{m^2} \quad \text{and} \quad x - \alpha = -\frac{2y^2}{m} - \frac{m}{2}:$$

substituting for y its value in the equation of the involute

$$y = m^{\frac{1}{2}} x^{\frac{1}{2}},$$

we obtain

$$-\beta = \frac{4x^{\frac{3}{2}}}{m^{\frac{1}{2}}}, \qquad x - \alpha = -2x - \frac{m}{2};$$

and by eliminating x, we have

$$\beta^2 = \frac{16}{27m}\left(\alpha - \frac{1}{2}m\right)^3,$$

which is the equation of the evolute.

If we make $\beta = 0$, we have

$$\alpha = \frac{1}{2}m;$$

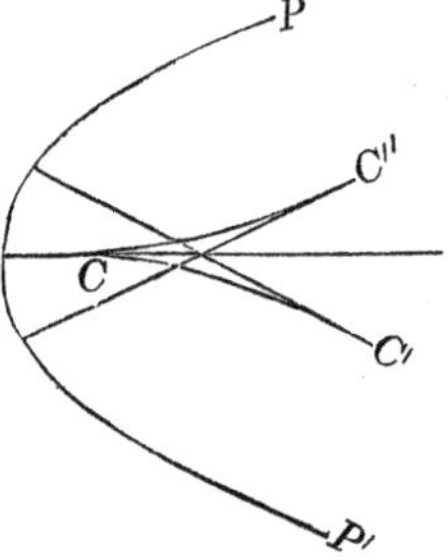

and hence, the evolute meets the axis of abscissas at a distance from the origin equal to half the parameter. If the origin of co-ordinates be transferred from A to this point, we shall have

$$\alpha' = \alpha - \frac{1}{2}m,$$

and consequently

$$\beta^2 = \frac{16}{27m}\alpha'^3.$$

The equation of the curve shows that it is symmetrical with respect to the axis of abscissas, and that it does not extend in the direction of the negative values of α'. The evolute CC' corresponds to the part AP of the involute, and CC'' to the part AP'.

CHAPTER VIII.

Of Transcendental Curves.—Of Tangent Planes and Normal Lines to Surfaces.

174. Curves may be divided into two general classes:

1st. Those whose equations are purely algebraic; and

2dly. Those whose equations involve transcendental quantities.

The first class are called algebraic curves, and the second, *transcendental curves.*

The properties of the first class having been already examined, it only remains to discuss the properties of the transcendental curves.

Of the Logarithmic Curve.

175. The logarithmic curve takes its name from the property that, when referred to rectangular axes, one of the co-ordinates is equal to the logarithm of the other.

If we suppose the logarithms to be estimated in parallels to the axis of Y, and the corresponding numbers to be laid off on the axis of abscissas, the equation of the curve will be

$$y = lx.$$

176. If we designate the base of a system of logarithms by a, we shall have, (Alg. Art. 241)

$$a^y = x;$$

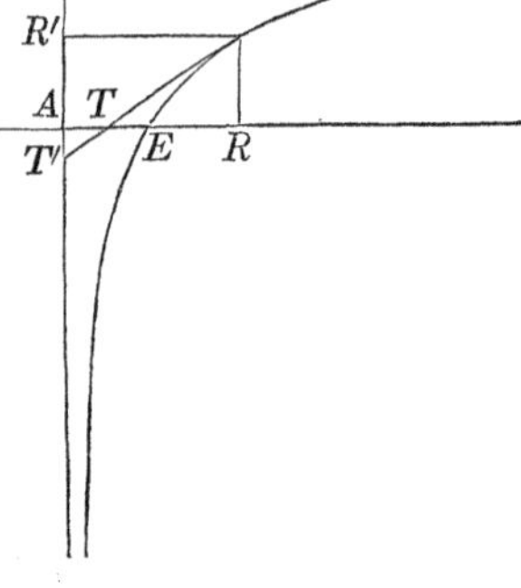

and if we change the value of the base a to a', we shall have

$$a'^y = x.$$

It is plain, that the same value of x, in the two equations, will give different values of y, and hence, *every system of logarithms will give a different logarithmic curve.*

If we make $y = 0$, we shall have (Alg. Art. 257) $x = 1$; and this relation being independent of the base of the system of logarithms, it follows, that *every logarithmic curve will intersect the axis of numbers at a distance from the origin equal to unity.*

The equation

$$a^y = x,$$

will enable us to describe the curve by points, even without the aid of a table of logarithms. For, if we make

$$y = 0, \quad y = \frac{1}{2}, \quad y = \frac{3}{2}, \quad y = \frac{1}{4}, \quad \text{\&c.},$$

we shall find, for the corresponding values of x,

$$x = 1, \quad x = \sqrt{a}, \quad x = a\sqrt{a}, \quad x = \sqrt[4]{a} \quad \text{\&c.}$$

177. If we suppose the base of the system of logarithms to be greater than unity, the logarithms of all numbers less

than unity will be negative (Alg. Art. 256); and therefore, the values of y corresponding to the abscissas, between the limits $x=0$ and $x=AE=1$, will be negative. Hence, these ordinates are laid off below the axis of abscissas.

When $x=0$, y will be infinite and negative (Alg. Art. 264). If we make x negative, the conditions of the equation cannot be fulfilled; and hence, the curve does not extend on the side of the negative abscissas.

178. Let us resume the equation of the curve

$$y = lx.$$

If we represent the modulus of the system of logarithms by A, and differentiate, we obtain (Art. 56),

$$dy = A\frac{dx}{x},$$

or

$$\frac{dy}{dx} = \frac{A}{x}.$$

But $\frac{dy}{dx}$ represents the tangent of the angle which the tangent line forms with the axis of abscissas: hence, the tangent will be parallel to the axis of abscissas when $x=\infty$, and perpendicular to it when $x=0$.

But when $x=0$, $y=-\infty$; hence, the axis of ordinates is an asymptote to the curve. The tangent which is parallel to the axis of X is not an asymptote: for when $x=\infty$, we also have $y=\infty$.

179. The most remarkable property of this curve belongs to its sub-tangent $T'R'$, estimated on the axis of logarithms. We have found, for the sub-tangent, on the axis of X (Art. 114),

$$TR = \frac{dx}{dy}y,$$

and by simply changing the axes, we have

$$T'R' = \frac{dy}{dx}x = A:$$

hence, *the sub-tangent is equal to the modulus of the system of logarithms from which the curve is constructed.* In the Naperian system $M = 1$, and hence the sub-tangent will be equal to $1 = AE$.

Of the Cycloid.

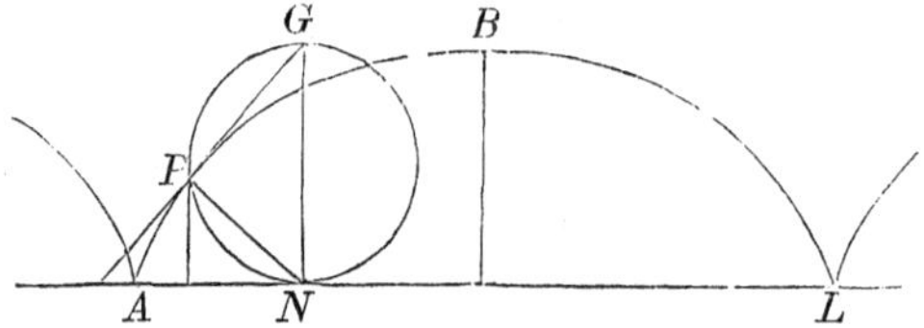

180. If a circle NPG be rolled along a straight line AL, any point of the circumference will describe a curve, which is called a *cycloid.* The circle NPG is called the *generating circle*, and P the *generating point.*

It is plain, that in each revolution of the generating circle an equal curve will be described; and hence, it will only be necessary to examine the properties of the curve $APBL$, described in one revolution of the generating circle. We shall therefore refer only to this part when speaking of the cycloid.

181. If we suppose the point P to be on the line AL at A, it will be found at some point, as L, after all the

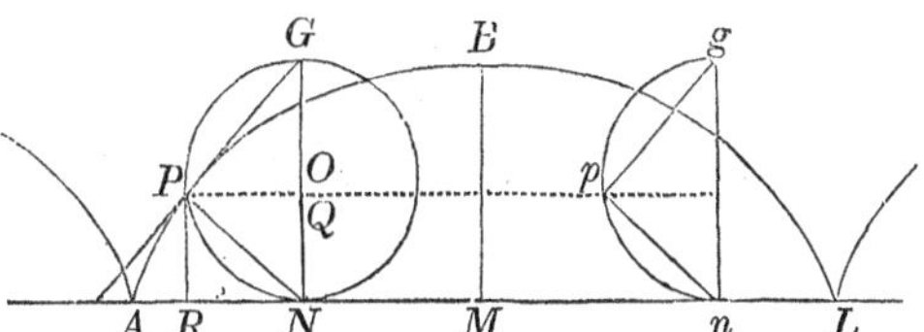

points of the circumference shall have been brought in contact with the line AL. The line AL will be equal to the circumference of the generating circle, and is called the *base of the cycloid.* The line BM, drawn perpendicular to the base at the middle point, is equal to the diameter of the generating circle, and is called the *axis of the cycloid.*

182. To find the equation of the cycloid, let us assume the point A as the origin of co-ordinates, and let us suppose that the generating point has described the arc AP. If N designates the point at which the generating circle touches the base, AN will be equal to the arc NP.

Through N draw the diameter NG, which will be perpendicular to the base. Through P draw PR perpendicular to the base, and PQ parallel to it. Then, $PR=NQ$ will be the versed-sine, and PQ the sine of the arc NP.

Let us make

$$ON=r, \qquad AR=x, \qquad PR=NQ=y,$$

we shall then have

$$PQ=\sqrt{2ry-y^2}, \quad x=AN-RN=\text{arc}\,NP-PQ:$$

hence, the transcendental equation is

$$x=\text{ver-sin}^{-1}y-\sqrt{2ry-y^2}.$$

183. The properties of the cycloid are, however, most easily deduced from its differential equation, which is readily found by differentiating both members of the transscendental equation.

We have (Art. 71),

$$d(\text{ver-sin}^{-1}y) = \frac{rdy}{\sqrt{2ry - y^2}},$$

$$d\left(-\sqrt{2ry - y^2}\right) = -\frac{rdy - ydy}{\sqrt{2ry - y^2}}:$$

hence,

$$dx = \frac{rdy}{\sqrt{2ry - y^2}} - \frac{rdy - ydy}{\sqrt{2ry - y^2}},$$

or

$$dx = \frac{ydy}{\sqrt{2ry - y^2}};$$

which is the differential equation of the cycloid.

184. If we substitute in the general equations of (Arts. 114, 115, 116, 117), the values of dx, dy, deduced from the differential equation of the cycloid, we shall obtain the values of the normal, sub-normal, tangent, and sub-tangent. They are,

$$\text{normal } PN = \sqrt{2ry}, \qquad \text{sub-normal } RN = \sqrt{2ry - y^2},$$

$$\text{tangent } PT = \frac{y\sqrt{2ry}}{\sqrt{2ry - y^2}}, \quad \text{sub-tangent } TR = \frac{y^2}{\sqrt{2ry - y^2}}.$$

These values are easily constructed, in consequence of their connexion with the parts of the generating circle.

The sub-normal RN, for example, is equal to PQ of the generating circle, since each is equal to $\sqrt{2ry - y^2}$: hence, the normal PN and the diameter GN intersect the base of the cycloid at the same point.

Now, since the tangent to the cycloid at the point P is perpendicular to the normal, it must coincide with the chord PG of the generating circle.

If, therefore, it be required to draw a normal or a tangent to the cycloid, at any point as P, draw any line, as ng, perpendicular to the base AL, and make it equal to the diameter of the generating circle. On ng describe a semi-circumference, and through P draw a parallel to the base of the cycloid. Through p, where the parallel cuts the semi-circumference, draw the supplementary chords pn, pg, and then draw through P the parallels PN, PG, and PN will be a normal, and PG a tangent to the cycloid at the point P.

185. Let us resume the differential equation of the cycloid

$$dx = \frac{ydy}{\sqrt{2ry - y^2}},$$

which may be put under the form

$$\frac{dy}{dx} = \frac{\sqrt{2ry - y^2}}{y} = \sqrt{\frac{2r}{y} - 1}.$$

If we make $y = 0$, we shall have

$$\frac{dy}{dx} = \infty;$$

and if we make $y = 2r$, we shall have

$$\frac{dy}{dx} = 0:$$

hence, the tangent lines drawn to the cycloid at the points where the curve meets the base, are perpendicular to the base; and the tangent drawn through the extremity of the greatest ordinate, is parallel to the base.

186. If we differentiate the equation

$$dx = \frac{ydy}{\sqrt{2ry - y^2}},$$

regarding dx as constant, we obtain

$$0 = (yd^2y + dy^2)\sqrt{2ry - y^2} - \frac{ydy(rdy - ydy)}{\sqrt{2ry - y^2}};$$

or by reducing and dividing by y,

$$0 = (2ry - y^2)d^2y + rdy^2,$$

whence we obtain

$$d^2y = -\frac{rdy^2}{2ry - y^2};$$

and hence the cycloid is concave towards the axis of abscissas (Art. 133).

187. To find the evolute of the cycloid, let us first substitute in the general value of

$$R = -\frac{(dx^2 + dy^2)^{\frac{3}{2}}}{dx\,d^2y},$$

the value of d^2y found in the last article: we shall then have

$$R = 2^{\frac{3}{2}}(ry)^{\frac{1}{2}} = 2\sqrt{2ry}:$$

hence, the radius of curvature corresponding to the extremity of any ordinate y, is equal to double the normal.

The radius of curvature is 0 when $y=0$, and equal to twice the diameter of the generating circle for $y=2r$: hence, the length of the evolute curve from A to A' is equal to twice the diameter of the generating circle.

Substituting the value of d^2y in the values of $y-\beta$, $x-\alpha$ (Art. 172), we obtain

$$y-\beta=2y, \qquad x-\alpha=-2\sqrt{2ry-y^2};$$

hence we have

$$y=-\beta, \qquad x=\alpha-2\sqrt{-2r\beta-\beta^2}.$$

Substituting these values of y and x in the transcendental equation of the cycloid, we have

$$\alpha=\text{ver-sin}^{-1}-\beta+\sqrt{-2r\beta-\beta^2},$$

which is the transcendental equation of the evolute, referred to the primitive origin and the primitive axes.

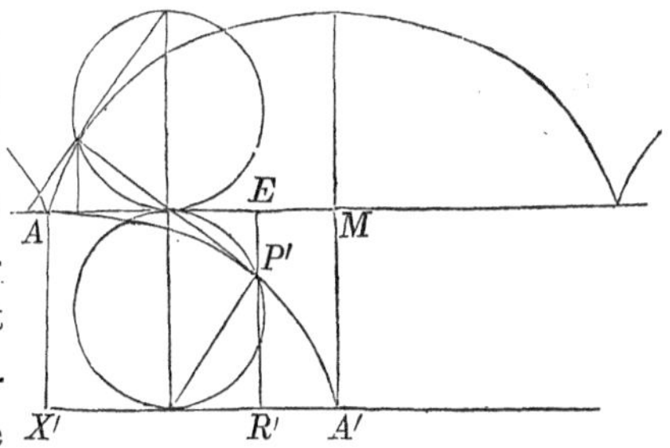

Let us now transfer the origin of co-ordinates to the point A', and change at the same time the direction of the positive abscissas: that is, instead of estimating them from the left to the right, we will estimate them from the right to the left. Let us designate the co-ordinates of the evolute, referred to the new axes $A'M$, $A'X'$, by α' and β'

Since $A'X' = AM$ = the semi-circumference of the generating circle, which is equal to $r\pi$, we shall have, for the abscissa $A'R'$ of any point P',

$$A'R' = \alpha' = r\pi - \alpha, \qquad \text{hence}, \qquad \alpha = r\pi - \alpha':$$

and for the ordinate, we shall have

$$R'P' = \beta' = R'E - P'E = 2r - (-\beta) = 2r + \beta,$$

hence, $\qquad \beta = -2r + \beta', \qquad$ or $\qquad -\beta = 2r - \beta'.$

Substituting these values of α and β in the transcen dental equation of the evolute, we obtain

$$r\pi - \alpha' = \text{ver-sin}^{-1}(2r - \beta') + \sqrt{2r\beta' - \beta'^2},$$

or $\qquad \alpha' = r\pi - \text{ver-sin}^{-1}(2r - \beta') - \sqrt{2r\beta' - \beta'^2}.$

But the arc whose versed-sine is $2r - \beta'$, is the supple ment of the arc whose versed-sine is β', hence

$$\alpha' = \text{ver-sin}^{-1}\ \beta' - \sqrt{2r\beta' - \beta'^2},$$

which is the equation of the evolute referred to the new origin and new axes.

But this equation is of the same form, and involves the same constants as that of the involute: hence, the evolute and involute are equal curves.

Of Spirals.

188. *A spiral* is a curve described by a point which moves along a right line, according to any law whatever, the line having at the same time a uniform angular motion.

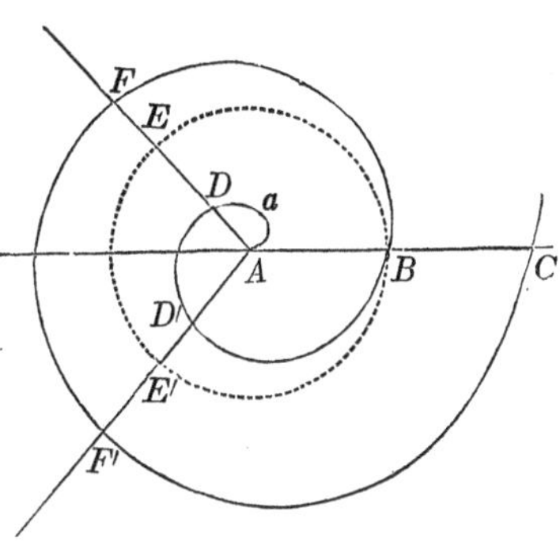

Let ABC be a straight line which is to be turned uniformly around the point A. When the motion of the line begins, let us suppose a point to move from A along the line in the direction ABC. When the line takes the position ADE the point will have moved along it to some point as D, and will have described the arc AaD of the spiral. When the line takes the position $AD'E'$ the point will have described the curve $AaDD'$, and when the line shall have completed an entire revolution the point will have described the curve $AaDD'B$.

The point A, about which the right line moves, is called the *pole*; the distances AD, AD', AB, are called *radius-vectors*, and if the revolutions of the radius-vector are continued, the generating point will describe an indefinite spiral. The parts $AaDD'B$, $BFF'C$, described in each revolution, are called *spires*.

189. If with the pole as a centre, and AB, the distance passed over by the generating point in the direction of the radius-vector during the first revolution, as a radius, we describe the circumference BEE', the angular motion of the radius-vector about the pole A, may be measured by the arcs of this circle, estimated from B.

If we designate the radius-vector by u, and the measuring arc, estimated from B, by t, the relation between u

and t, may in general be expressed by the equation

$$u = at^n,$$

in which n depends on the *law* according to which the generating point moves along the radius-vector, and a on the relation which exists between a given value of u and the corresponding value of t.

190. When n is positive the spirals represented by the equation

$$u = at^n,$$

will pass through the pole A. For, if we make $t = 0$, we shall have $u = 0$.

But if n is negative, the equation will become

$$u = at^{-n}, \qquad \text{or} \qquad u = \frac{a}{t^n},$$

in which we shall have

$$u = \infty \qquad \text{for} \qquad t = 0,$$

and

$$u = 0 \qquad \text{for} \qquad t = \infty:$$

hence, in this class of spirals, the first position of the generating point is at an infinite distance from the pole: the point will then approach the pole as the radius-vector revolves, and will only reach it after an infinite number of revolutions.

191. If we make $n = 1$, the equation of the spiral becomes

$$u = at.$$

If we designate two different radius-vectors by u' and u'', and the corresponding arcs by t' and t'', we shall have

$$u' = at', \qquad \text{and} \qquad u'' = at'',$$

and consequently

$$u' : u'' :: t' : t'';$$

that is, *the radius-vectors are proportional to the measuring arcs, estimated from the point B.* This spiral is called, the spiral of Archimedes.

192. If we represent by unity the distance which the generating point moves along the radius-vector, during one revolution, the equation

$$u = at,$$

will become

$$1 = at, \qquad \text{or} \qquad 1 \times \frac{1}{a} = t.$$

But since t is the circumference of a circle whose radius is unity, we shall have

$$\frac{1}{a} = 2\pi, \qquad \text{and consequently,} \qquad a = \frac{1}{2\pi}.$$

193. If the axis BD, of a semi-parabola BCD, be wrapped around the circumference of a circle of a given radius r, any abscissa, as Bb, will coincide with an equal arc Bb', and any ordinate as ba, will take the direction of the normal $Ab'a'$. The curve $Ba'c'$, described through the extremities of the ordinates of the parabola, is called the *parabolic spiral.*

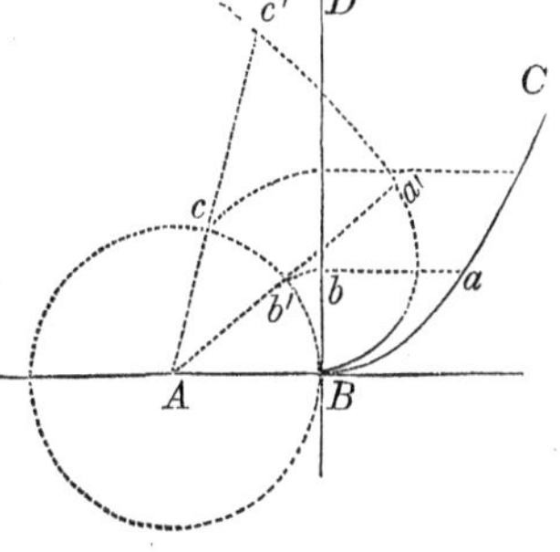

The equation of this spiral is readily found, by observing that the squares of the lines $b'a'$, cc', &c., are proportional to the abscissas or arcs Bb', Bc.

If we designate the distances, estimated from the pole A, by u, we shall have $b'a' = u - r$: hence,

$$(u - r)^2 = 2pt,$$

is the equation of the parabolic spiral.

If we suppose $r = 0$, the equation becomes

$$u^2 = 2pt.$$

If we make $n = -1$, the general equation of spirals becomes

$$u = at^{-1}, \qquad \text{or} \qquad ut = a.$$

This spiral is called the *hyperbolic spiral*, because of the analogy which its equation bears to that of the hyperbola, when referred to its asymptotes.

194. The relation between u and t is entirely arbitrary, and besides the relations expressed by the equation

$$u = at^n,$$

we may, if we please, make

$$t = \log u.$$

The spiral described by the extremity of the radius-vector when this relation subsists, is called the *logarithmic spiral*.

195. If in the equation of the hyperbolic spiral, we make successively,

$$t = 1, \quad = \frac{1}{2}, \quad = \frac{1}{3}, \quad = \frac{1}{4}, \text{ \&c.},$$

we shall have the corresponding values,

$$u = a, \quad u = 2a, \quad u = 3a, \quad u = 4a, \text{ \&c.}$$

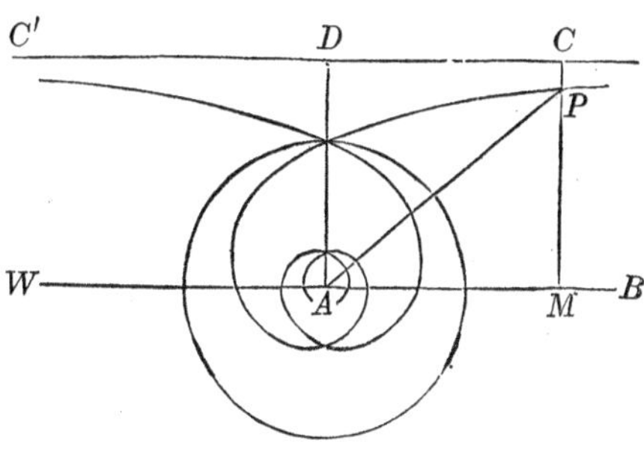

Through the pole A draw AD perpendicular to AB, and make it equal to a: then through D draw a parallel to AB. From any point of the spiral as P draw PM perpendicular to AB, we shall then have

$$PM = u \sin MAP = u \sin t.$$

If we substitute for u its value $\frac{a}{t}$, we shall have

$$PM = a\frac{\sin t}{t}.$$

Now as the arc t diminishes, the ratio of $\frac{\sin t}{t}$ will approach to unity, and the value of the ordinate PM will approach to a or CM: hence, the line DC approaches the curve and becomes tangent to it when $t = 0$. But when $t = 0$, $u = \infty$; hence, the line DC is an asymptote of the curve.

196. The arc which measures the angular motion of the radius-vector has been estimated from the right to the left, and the value of t regarded as positive. If we revolve the radius-vector in a contrary direction, the measuring arc will be estimated from left to right, the sign of t will be changed to negative and a similar spiral will be described. The line DC' is an asymptote to the hyperbolic spiral, corresponding to the negative value of t.

197. Let us now find a general value for the subtangent of any curve referred to polar co-ordinates. *The subtangent is the projection of the tangent on a line drawn through the pole and perpendicular to the radius-vector passing through the point of contact.*

The equation of the curve may be written under the form

$$u = f(t),$$

in which we may suppose t the independent variable, and its first differential constant.

Let $AO = 1$ be the radius of the measuring circle, PT a tangent to the curve at the point P, and AT drawn perpendicular to the radius-vector AP, the subtangent.

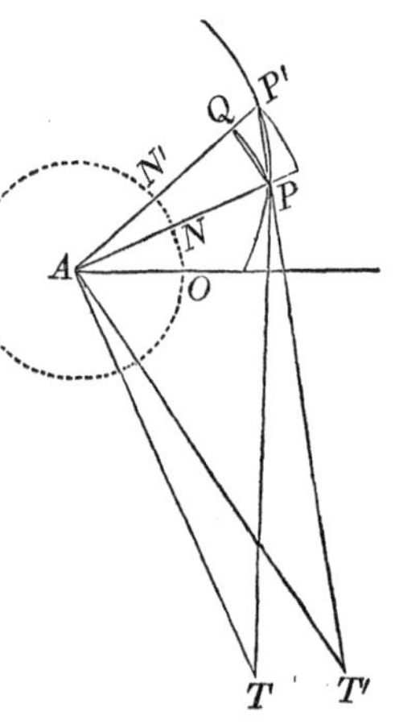

Take any other point of the curve as P', and draw AP'. About the centre A describe the arc PQ, and draw the chord PQ. Draw also the secant PP' and prolong it until it meets AT', drawn parallel to QP, at T'.

From the similar triangles QPP', $AT'P'$, we have

$$PQ : QP' :: AT' : AP';$$

hence,

$$\frac{QP'}{PQ} = \frac{AP'}{AT'}.$$

But when we pass to the limit, by supposing the point P' to coincide with P, the secant $T'PP'$ will become the tangent PT, and AT' will become the subtangent AT.

But under this supposition the arc NN' will become equal to dt, the arc PQ to the chord PQ (Art. 128), AP' to u, and the line QP' to du.

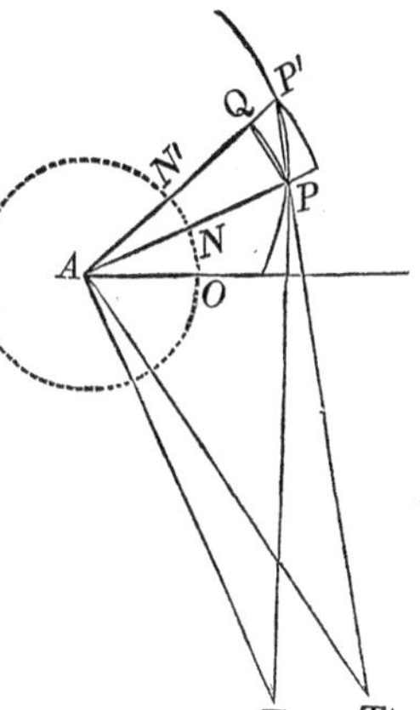

To find the value of the arc PQ, we have

$$1 : NN' :: AP : \text{arc } PQ;$$

hence,

$$1 : dt :: u : \text{arc } PQ,$$

and $$PQ = udt.$$

Substituting these values, and passing to the limit, we have

$$\frac{du}{udt} = \frac{u}{AT}:$$

hence, we have the subtangent

$$AT = \frac{u^2 dt}{du}.$$

198. If we find the value of u^2 and du from the general equation of the spirals

$$u = at^n,$$

we shall have

$$AT = \frac{a}{n} t^{n+1}.$$

In the spiral of Archimedes, we have

$$n=1, \quad \text{and} \quad a=\frac{1}{2\pi};$$

hence,
$$AT=\frac{t^2}{2\pi}.$$

If now we make $t=2\pi=$ circumference of the measuring circle, we shall have

$$AT=2\pi = \text{circumference of measuring circle.}$$

After m revolutions, we shall have

$$t=2m\pi,$$

and consequently,

$$AT=2m^2\pi = m.2m\pi;$$

that is, *the subtangent, after* m *revolutions, is equal to* m *times the circumference of the circle described with the radius-vector.* This property was discovered by Archimedes.

199. In the hyperbolic spiral $n=-1$, and the value of the subtangent becomes

$$AT=-a;$$

that is, the subtangent is constant in the hyperbolic spiral.

200. It may be remarked, that

$$\frac{AT}{AP}=\frac{udt}{du}$$

expresses the tangent of the angle which the tangent makes with the radius-vector.

In the logarithmic spiral, of which the equation is

$$t = \log u,$$

we have
$$dt = A\frac{du}{u};$$

hence,
$$\frac{AT}{AP} = \frac{udt}{du} = A;$$

that is, in the logarithmic spiral, the angle formed by the tangent and the radius-vector passing through the point of contact, is constant; and the tangent of the angle is equal to the modulus of the system of logarithms. If t is the Naperian logarithm of u, the angle will be equal to $45°$.

201. The value of the tangent in a curve referred to polar co-ordinates,

$$PT = \sqrt{\overline{AP}^2 + \overline{AT}^2} = u\sqrt{1 + \frac{u^2dt^2}{du^2}}.$$

202. To find the differential of the arc, which we will represent by z, we have

$$PP' = \sqrt{\overline{QP'}^2 + \overline{QP}^2};$$

or, by substituting for QP' and PQ their values, and passing to the limit, we have

$$dz = \sqrt{du^2 + u^2dt^2}.$$

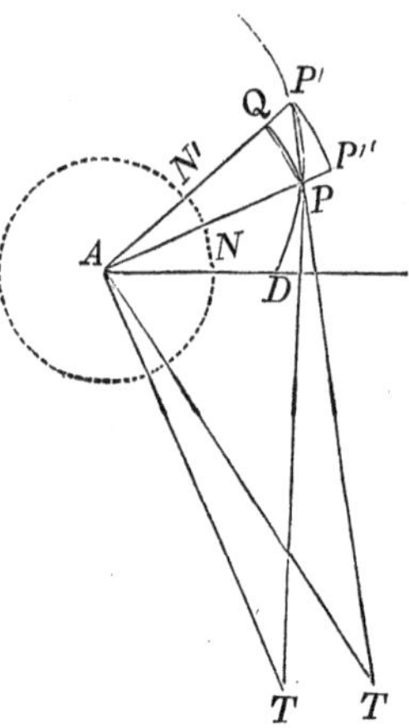

203. The differential of the area ADP when referred to the polar co-ordinates, is not an elementary rectangle as when referred to rectangular axes, but is the elementary sector APP'.

The limit of the ratio of the sector APP' with the arc NN', will be the same as that of either of the sectors APQ, $AP''P'$ between which it is contained, with the same arc NN'. Hence, if we designate the area by s, and pass to the limit, we shall have

$$\frac{ds}{dt} = \frac{AP \times PQ}{2NN'} = \frac{u^2}{2} \quad \text{or} \quad ds = \frac{u^2 dt}{2};$$

which is the differential of the area of any segment of a spiral.

Of Tangent Planes and Normal Lines to Surfaces.

204. Let $$u = F(x, y, z) = 0,$$

be the equation of a surface.

If through any point of the surface two planes be passed intersecting the surface in two curves, and two straight lines be drawn respectively tangent to each of the curves, at their common point, the plane of these tangents will be tangent to the surface.

205. Let us designate the co-ordinates of the point at which the plane is to be tangent by x'', y'', z''.

Through this point let a plane be passed parallel to the co-ordinate plane YZ. This plane will intersect the surface in a curve. The equations of a straight line tangent to this curve, at the point whose co-ordinates are x'', y'', z'', are

$$x = x'' = a'', \qquad y - y'' = \frac{dy''}{dz''}(z - z'');$$

the first equation represents the projection of the tangent on the co-ordinate plane ZX, and the second its projection on the co-ordinate plane YZ (An. Geom. Bk. IX. Art. 70).

Through the same point let a plane be passed parallel to the co-ordinate plane ZX, and we shall have for the equations of a tangent to the curve

$$y = y'' = b'', \qquad x - x'' = \frac{dx''}{dz''}(z - z'');$$

The coefficient $\frac{dy}{dz}$ represents the tangent of the angle which the projection of the first tangent on the co-ordinate plane YZ makes with the axis of Z; and the coefficient $\frac{dx}{dz}$ represents the tangent of the angle which the projection of the second tangent on the plane ZX makes with the axis of Z (An. Geom. Bk. VIII, Prop. II).

But these coefficients may be expressed in functions of the surface and the co-ordinates of its points. For, we have

$$u = f(x, y, z) = 0,$$

and if we suppose x constant, we shall have (Art. 87)

$$du = \frac{du}{dy}dy + \frac{du}{dz}dz = 0:$$

hence,
$$\frac{dy}{dz} = -\frac{\frac{du}{dz}}{\frac{du}{dy}}:$$

and if we suppose y constant, we shall find, in a similar manner,

$$\frac{dx}{dz} = -\frac{\frac{du}{dz}}{\frac{du}{dx}};$$

hence, the equation of the projection of the first tangent on the plane of YZ becomes

$$y - y'' = -\frac{\frac{du}{dz}}{\frac{du}{dy}}(z - z'');$$

and the equation of the projection of the second tangent on the plane of ZX is

$$x - x'' = -\frac{\frac{du}{dz}}{\frac{du}{dx}}(z - z'').$$

The equation of a plane passing through the point whose co-ordinates are x'', y'', z'' is of the form

$$A(x - x'') + B(y - y'') + C(z - z'') = 0,$$

in which $-\frac{C}{B}$ will represent the tangent of the angle which the trace on the co-ordinate plane YZ makes with the axis of Z, and $-\frac{C}{A}$ the tangent of the angle which the trace on the plane of ZX makes with the axis of Z.

But since the tangents are respectively parallel to the co-ordinate planes YZ, ZX, their projections will be parallel to the traces of the tangent plane: therefore,

$$-\frac{C}{B} = -\frac{\frac{du}{dz}}{\frac{du}{dy}}, \quad \text{hence,} -B = -\frac{\frac{du}{dy}}{\frac{du}{dz}}C;$$

$$-\frac{C}{A} = -\frac{\frac{du}{dz}}{\frac{du}{dx}}, \quad \text{hence,} -A = -\frac{\frac{du}{dx}}{\frac{du}{dz}}C.$$

Substituting these values of B and A in the equation of the plane, and reducing, we find

$$(z - z'')\frac{du}{dz} + (x - x'')\frac{du}{dx} + (y - y'')\frac{du}{dy} = 0,$$

which is the equation of a tangent plane to a surface at a point of which the co-ordinates are x'', y'', z''.

206. A normal line to the surface being perpendicular to the tangent plane at the point of contact, its equations will be of the form

$$x - x'' = \frac{\frac{du}{dx}}{\frac{du}{dz}}(z - z''), \qquad y - y = \frac{\frac{du}{dy}}{\frac{du}{dz}}(z - z'').$$

ELEMENTS

OF THE

INTEGRAL CALCULUS.

Integration of Differential Monomials.

207. The Differential Calculus explains the method of finding the differential of a given function. The Integral Calculus is the reverse of this. It explains the method of finding the function which corresponds to a given differential.

The rules for the differentiation of functions are explicit and direct. Those for determining the integral, or function, from the differential expression, are less direct and are deduced by reversing the process by which we pass from the function to the differential.

208. Let it be required, as a first example, to integrate the expression.

$$x^m dx.$$

We have found (Art. 32), that

$$d(x^{m+1}) = (m+1)x^m dx,$$

whence, $$x^m dx = \frac{dx^{m+1}}{m+1} = d\left(\frac{x^{m+1}}{m+1}\right),$$

and consequently $$\frac{x^{m+1}}{m+1},$$

is the function of which the differential is $x^m dx$.

The integration is indicated by placing the character $\int$ before the differential which is to be integrated. Thus, we write

$$\int x^m dx = \frac{x^{m+1}}{m+1},$$

from which we deduce the following rule.

To integrate a monomial of the form $x^m dx$, *augment the exponent of the variable by unity, and divide by the exponent so increased and by the differential of the variable.*

209. The characteristic $\int$ signifies *integral* or *sum.* The word *sum,* was employed by those who first used the differential and integral calculus, and who regarded the integral of

$$x^m dx$$

as the *sum* of all the products which arise by multiplying the *m*th power of x, for all values of x, by the constant dx.

210. Let it be required to integrate the expression $\frac{dx}{x^3}$. We have, from the last rule,

$$\int \frac{dx}{x^3} = \int dx\, x^{-3} = \frac{x^{-3+1}}{-3+1} = \frac{x^{-2}}{-2} = -\frac{1}{2x^2}.$$

In a similar manner, we find

$$\int dx \sqrt[3]{x^2} = \int x^{\frac{2}{3}} dx = \frac{x^{\frac{2}{3}+1}}{\frac{2}{3}+1} = \frac{x^{\frac{5}{3}}}{\frac{5}{3}} = \frac{3x^{\frac{5}{3}}}{5}.$$

211. It has been shown (Art. 22), that the differential of the product of a variable multiplied by a constant, is equal to the constant multiplied by the differential of the variable. Hence, we may conclude that, *the integral of the product of a differential by a constant, is equal to the constant multiplied by the integral of the differential:* that is,

$$\int ax^m dx = a\int x^m dx = a\frac{x^{m+1}}{m+1}.$$

Hence, *if the expression to be integrated have one or more constant factors, they may be placed as factors without the sign of the integral.*

212. It has also been shown (Art. 22), that every constant quantity connected with the variable by the sign plus or minus, will disappear in the differentiation; and hence, the differential of $a+x^m$, is the same as that of x^m; viz. $mx^{m-1}dx$. Consequently, the same differential may answer to several integral functions differing from each other in the value of the constant term.

In passing, therefore, from the differential to the integral or function, we must annex to the first integral obtained, a constant term, and then find such a value for this term as will characterize the particular integral sought.

For example (Art. 94),

$$\frac{dy}{dx}=a, \quad \text{or} \quad dy=adx,$$

is the differential equation of every straight line which makes with the axis of abscissas an angle whose tangent is a. Integrating this expression, we have

$$\int dy = a \int dx,$$

or $$y = ax,$$

or finally, $$y = ax + C.$$

If now, the required line is to pass through the origin of co-ordinates, we shall have, for

$$x = 0, \quad y = 0, \quad \text{and consequently,} \quad C = 0.$$

But if it be required that the line shall intersect the axis of Y at a distance from the origin equal to $+b$, we shall have, for

$$x = 0, \quad y = +b, \quad \text{and consequently,} \quad C = +b;$$

and the true integral will be

$$y = ax + b.$$

If, on the contrary, it were required that the right line should intersect the axis of ordinates below the origin, we should have, for

$$x = 0, \quad y = -b, \quad \text{and consequently,} \quad C = -b;$$

and the true integral would be

$$y = ax - b.$$

213. It has been shown (Art. 95), that

$$xdx + ydy = 0$$

is the differential equation of the circumference of a circle By taking the integral, we have

$$\int xdx + \int ydy = 0, \quad \text{or} \quad x^2 + y^2 = 0,$$

or finally, $$x^2 + y^2 + C = 0.$$

If it be required that this integral shall represent a given circumference, of which the radius is R, we shall have, by making

$$x=0, \qquad y^2=-C=R^2,$$

and hence, $$C=-R^2;$$

and consequently the true integral is

$$x^2+y^2-R^2=0, \qquad \text{or} \qquad x^2+y^2=R^2.$$

The constant C, which is annexed to the first integral that is obtained, is called an *arbitrary constant*, because such a value is to be attributed to it as will cause the required integral to fulfil given conditions, which may be imposed on it at pleasure.

The value of the constant must be such, as to *render the equation true for every value which can be attributed to the variables.*

214. There is one case to which the formula of Art. 208 does not apply. It is that in which $m=-1$. Under this supposition,

$$\int x^m dx=\frac{x^{m+1}}{m+1}=\frac{x^{-1+1}}{-1+1}=\frac{x^0}{0}=\frac{1}{0}=\infty.$$

But when $m=-1$,

$$\int x^m dx=\int x^{-1}dx=\int\frac{dx}{x},$$

and $$\int\frac{dx}{x}=\log x+C. \quad \text{(Art. 57)}.$$

215. Since the differential of a function composed of several terms, is equal to the sum or difference of the differentials (Art. 27), it follows that the integral of a differen-

tial expression, composed of several terms, is equal to the sum or difference of the integrals taken separately. For example, if

$$du = adx - \frac{bdx}{x^3} + x\sqrt{x}\,dx, \qquad \text{we have}$$

$$\int du = \int (adx - \frac{bdx}{x^3} + x\sqrt{x}\,dx), \qquad \text{and}$$

$$u = ax + \frac{1}{2}\frac{b}{x^2} + \frac{2}{5}x^{\frac{5}{2}} + C.$$

216. Every polynomial of the form

$$(a + bx + cx^2 + \&c.)^n dx,$$

in which n is a positive and whole number, may be integrated by the rule for monomials, by first raising the polynomial to the power indicated by the exponent, and then multiplying each term by dx.

If, for example, we make $n = 2$, and employ but two terms, we have

$$\int (a + bx)^2 dx = \int (a^2dx + 2\,abx\,dx + b^2x^2dx),$$

$$= a^2x + abx^2 + \frac{b^2x^3}{3} + C.$$

Integration of Particular Binomials.

217. If we have a binomial of the form

$$du = (a + bx^n)^m x^{n-1} dx;$$

that is, *in which the exponent of the variable without the parenthesis is less by unity than the exponent of the variable within*, we may make

$$a + bx^n = z, \quad \text{which gives}$$

$$nbx^{n-1}dx = dz, \qquad \text{or} \qquad x^{n-1}dx = \frac{dz}{nb};$$

whence $\quad du = z^m \dfrac{dz}{nb}, \qquad$ or $\qquad u = \dfrac{z^{m+1}}{(m+1)nb};$

and consequently

$$u = \frac{(a + bx^n)^{m+1}}{(m+1)nb} + C.$$

Hence, the integral of the above form, is equal to *the binomial factor with its exponent augmented by unity, divided by the exponent so increased, into the exponent of the variable within the parenthesis into the coefficient of the variable.*

For example,

$$\int(a + 3x^2)^3 xdx = \frac{(a + 3x^2)^4}{4.2.3} + C; \qquad \text{and}$$

$$\int(a + bx^2)^{\frac{1}{2}} mxdx = \frac{m}{3b}(a + bx^2)^{\frac{3}{2}} + C.$$

218. A transformation similar to that of the last article will enable us to integrate certain differentials corresponding to logarithmic functions. If we have an expression of the form

$$du = \frac{adx}{c + bx},$$

make $c + bx = z$, which gives $dx = \dfrac{dz}{b}$, and by substituting, we have

$$\int\frac{adx}{c + bx} = \int\frac{a}{b}\frac{dz}{z} = \frac{a}{b}\int\frac{dz}{z} = \frac{a}{b}\log z + C,$$

and by substituting for z its value

$$\int \frac{adx}{c+bx} = \frac{a}{b}\log(c+bx) + C.$$

In a similar manner, we should find

$$\int \frac{adx}{c-bx} = -\frac{a}{b}\log(a-bx) + C,$$

in which the integral is negative, since $d(-x) = -dx$.

We can find, in a similar manner, the integral of every fraction of which *the numerator is equal to the differential of the denominator, or equal to that differential multiplied by a constant.*

If, for example, we have

$$du = \frac{(b+2cx)mdx}{a+bx+cx^2};$$

make $a+bx+cx^2 = z$, which gives, $bdx + 2cxdx = dz$, and hence,

$$du = \frac{mdz}{z}, \quad \text{or} \quad u = m\log z,$$

and by substituting for z its value

$$u = m\log(a+bx+cx^2).$$

Of Differentials whose Integrals are expressed by the Circular Functions.

219. We have seen, Art. 71, that if x designates an arc and u the sine, to the radius unity, we shall have

$$dx = \frac{du}{\sqrt{1-u^2}}.$$

hence, $$\int \frac{du}{\sqrt{1-u^2}} = x + C;$$

or adopting the notation of Art. 72,

$$\int \frac{du}{\sqrt{1-u^2}} = \sin^{-1}u + C.$$

If the arc expressed in the second member of the equation be estimated from the beginning of the first quadrant, the sine will be 0, when the arc is 0, and we shall have, for $u = 0$

$$\int \frac{du}{\sqrt{1-u^2}} = 0, \quad \text{and consequently} \quad C = 0,$$

and under this supposition, the entire integral is

$$\int \frac{du}{\sqrt{1-u^2}} = \sin^{-1}u.$$

To give an example, showing the use of the arbitrary constant, let us suppose that the arc which is to be expressed by the second member of the equation, is to be estimated from the beginning of the second quadrant. This supposition will render

$$\int \frac{du}{\sqrt{1-u^2}} = 0 \quad \text{for} \quad u = 1.$$

But when $u = 1$, $\sin^{-1}u = \frac{1}{2}\pi$; hence,

$$\frac{1}{2}\pi + C = 0, \quad \text{or} \quad C = -\frac{1}{2}\pi:$$

and we have, for the entire integral, under this supposition,

$$\int \frac{du}{\sqrt{1-u^2}} = \sin^{-1}u - \frac{1}{2}\pi.$$

220. It frequently happens that we have expressions to integrate of the form

$$\frac{dz}{\sqrt{a^2-z^2}}.$$

Let us suppose, for a moment, that a is the radius of a circle, and z the sine of any arc of the circle; and that u is the sine of an arc containing an equal number of degrees in a circle whose radius is unity: we shall then have,

$$1 : u :: a : z;$$

hence, $u=\frac{z}{a}$, and $du=\frac{dz}{a}$;

and consequently,

$$\int\frac{du}{\sqrt{1-u^2}}=\int\left[\frac{\frac{dz}{a}}{\sqrt{1-\frac{z^2}{a^2}}}\right]=\int\frac{dz}{\sqrt{a^2-z^2}};$$

hence, $\int\frac{du}{\sqrt{1-u^2}}=\int\frac{dz}{\sqrt{a^2-z^2}}=\sin^{-1}\frac{z}{a}$:

the arc being still taken in a circle whose radius is unity.

221. We have seen (Art. 71), that if x designates an arc, and u the cosine, to the radius unity, we shall have

$$dx=-\frac{du}{\sqrt{1-u^2}};$$

hence, $\int-\frac{du}{\sqrt{1-u^2}}=x+C$;

or adopting the notation of Art. 72,

$$\int-\frac{du}{\sqrt{1-u^2}}=\cos^{-1}u+C.$$

If the arc be estimated from the beginning of the first quadrant, it will be equal to $\frac{1}{2}\pi$ for $u=0$; hence, the first member of the equation becomes equal to $\frac{1}{2}\pi$ when $u=0$. But under this supposition, $\cos^{-1}u=\frac{1}{2}\pi$: hence, $C=0$, and the entire integral is

$$\int -\frac{du}{\sqrt{1-u^2}}=\cos^{-1}u.$$

222. By a method analogous to that of Art. 220, we should find

$$\int -\frac{dz}{\sqrt{a^2-z^2}}=\cos^{-1}\frac{z}{a},$$

the arc being estimated to the radius unity.

223. We have seen (Art. 71), that if x represents an arc, and u its tangent, to the radius unity, we have

$$dx=\frac{du}{1+u^2};$$

hence, $$\int\frac{du}{1+u^2}=x+C:$$

or, adopting the notation of Art. 72,

$$\int\frac{du}{1+u^2}=\text{tang}^{-1}u+C.$$

If the arc is estimated from the beginning of the first quadrant, we shall have

$\text{tang}^{-1}=0$, when $\int\frac{du}{1+u^2}=0$; hence, $C=0$, and the entire integral is

$$\int \frac{du}{1+u^2} = \text{tang}^{-1} u.$$

224. To integrate expressions of the form

$$\frac{dz}{a^2 + z^2};$$

let us suppose for a moment that a is the radius of a circle, and z the tangent of any arc, and that u is the tangent of an arc containing an equal number of degrees in a circle whose radius is unity: we shall then have, as in (Art. 220),

$$1 : u :: a : z;$$

hence, $\quad u = \frac{z}{a}, \quad u^2 = \frac{z^2}{a^2}, \quad$ and $\quad du = \frac{dz}{a},$

and consequently,

$$\int \frac{du}{1+u^2} = a \int \frac{dz}{a^2 + z^2} = \text{tang}^{-1} \frac{z}{a};$$

hence, by dividing by a,

$$\int \frac{dz}{a^2 + z^2} = \frac{1}{a} \text{tang}^{-1} \frac{z}{a},$$

the arc being estimated to the radius unity.

225. We have seen (Art. 71), that if x represents an arc, and u the versed-sine, to the radius of unity, we have

$$dx = \frac{du}{\sqrt{2u - u^2}};$$

hence, $\quad \int \frac{du}{\sqrt{2u - u^2}} = x = \text{ver-sin}^{-1} u + C:$

and if the arc is estimated from the beginning of the first quadrant, $C=0$, and we shall have

$$\int \frac{du}{\sqrt{2u-u^2}} = \text{ver-sin}^{-1} u.$$

226. To integrate an expression of the form

$$\frac{dz}{\sqrt{2az-z^2}}.$$

Suppose, as before, a to be the radius of a circle, and we shall have (Art. 220),

$$u = \frac{z}{a}, \qquad du = \frac{dz}{a};$$

and consequently,

$$\int \frac{du}{\sqrt{2u-u^2}} = \int \frac{dz}{\sqrt{2az-z^2}} = \text{ver-sin}^{-1} \frac{z}{a}$$

to the radius unity.

Integration by Series.

227. Every expression of the form

$$X dx,$$

in which X is such a function of x, that it can be developed in the powers of x, may be integrated by series.

For, let us suppose

$$X = Ax^a + Bx^b + Cx^c + Dx^d + \&c., \quad \text{then,}$$

$$Xdx = Ax^a dx + Bx^b dx + Cx^c dx + Dx^d dx + \&c.,$$

$$\int Xdx = \frac{A}{a+1} x^{a+1} + \frac{B}{b+1} x^{b+1} + \frac{C}{c+1} x^{c+1} + \frac{D}{d+1} x^{d+1} + \&c.$$

Hence, the integration by series is effected by *developing the function* X *in the powers of* x, *multiplying the series by* dx, *and then integrating the terms separately.*

Let us take, as a first example, $\dfrac{dx}{a+x}$,

$$\frac{dx}{a+x} = dx \times \frac{1}{a+x} = dx(a+x)^{-1},$$

$$(a+x)^{-1} = \frac{1}{a} - \frac{x}{a^2} + \frac{x^2}{a^3} - \frac{x^3}{a^4} + \text{\&c.};$$

and consequently,

$$\int \frac{dx}{a+x} = \int \left(\frac{1}{a}dx - \frac{xdx}{a^2} + \frac{x^2dx}{a^3} - \frac{x^3dx}{a^4} + \text{\&c.}\right):$$

and integrating each term separately, we obtain

$$\int \frac{dx}{a+x} = \frac{x}{a} - \frac{x^2}{2a^2} + \frac{x^3}{3a^3} - \frac{x^4}{4a^4} + \text{\&c.} + C,$$

and remarking that $\displaystyle\int \frac{dx}{a+x} = \log(a+x)$ (Art. 218), we have

$$\log(a+x) = \frac{x}{a} - \frac{x^2}{2a^2} + \frac{x^3}{3a^3} - \frac{x^4}{4a^4} + \text{\&c.} + C.$$

To determine the value of the constant, make $x=0$, which gives

$$\log a = 0 + C, \quad \text{or} \quad C = \log a; \quad \text{hence,}$$

$$\log(a+x) = \log a + \frac{x}{a} - \frac{x^2}{2a^2} + \frac{x^3}{3a^3} - \frac{x^4}{4a^4} + \text{\&c.},$$

$$\log(a+x) - \log a = \log\left(1+\frac{x}{a}\right) = \frac{x}{a} - \frac{x^2}{2a^2} + \frac{x^3}{3a^3} - \text{\&c.},$$

a result which agrees with the development in Art. 58.

228. Let us take, for a second example $\frac{dx}{1+x^2}$.

We have, $\frac{dx}{1+x^2} = dx(1+x^2)^{-1}$;

and by developing and integrating,

$$\int \frac{dx}{1+x^2} = x - \frac{x^3}{3} + \frac{x^5}{5} - \frac{x^7}{7} + \&c. + C.$$

When we make $x=0$, the arc is 0; hence,

$$\text{tang}^{-1}x = x - \frac{x^3}{3} + \frac{x^5}{5} - \frac{x^7}{7} + \&c.;$$

a result which corresponds with that of Art. 78.

229. If, in the expression $\frac{dx}{1+x^2}$, we place x^2 in the first term of the binomial, and then develop the binomial x^2+1, we obtain

$$\int \frac{dx}{x^2+1} = \int \left(\frac{1}{x^2} - \frac{1}{x^4} + \frac{1}{x^6} - \frac{1}{x^8} + \&c.\right) dx;$$

and by integrating, we have

$$\text{tang}^{-1}x = -\frac{1}{x} + \frac{1}{3x^3} - \frac{1}{5x^5} + \&c. + C.$$

To find the value of the constant C, let us make the arc $=90° = \frac{1}{2}\pi$. This supposition will render the tangent x infinite, and consequently every term of the series will become 0, and the equation will give

$$\frac{1}{2}\pi = 0 + C, \quad \text{or} \quad C = \frac{1}{2}\pi.$$

Making this substitution, we have, for the true integral,

$$\int \frac{dx}{x^2+1} = \text{tang}^{-1}x = \frac{1}{2}\pi - \frac{1}{x} + \frac{1}{3x^3} - \frac{1}{5x^5} + \&c.$$

230. The two series, found from the expressions $\frac{dx}{1+x^2}$ and $\frac{dx}{x^2+1}$, are, as they should be, essentially the same.

For, the tangent of an arc multiplied by its cotangent, is equal to radius square or unity (Trig. Art. XVIII). Hence, if we substitute for x, in the first series, $\frac{1}{x}$, we shall have, for the complemental arc,

$$\text{tang}^{-1}\frac{1}{x} = \frac{1}{x} - \frac{1}{3x^3} + \frac{1}{5x^5};$$

and subtracting both members from $\frac{1}{2}\pi$,

$$\frac{1}{2}\pi - \text{tang}^{-1}\frac{1}{x} = \text{tang}^{-1}x = \frac{1}{2}\pi - \frac{1}{x} + \frac{1}{3x^3} - \frac{1}{5x^5} + \&c.$$

231. We have found (Art. 71),

$$\sin^{-1}x = \int \frac{dx}{\sqrt{1-x^2}} = (1-x^2)^{-\frac{1}{2}}dx;$$

and by developing, we find

$$(1-x^2)^{-\frac{1}{2}} = 1 + \frac{1}{2}x^2 + \frac{1}{2}\cdot\frac{3}{4}x^4 + \frac{1}{2}\cdot\frac{3}{4}\cdot\frac{5}{6}x^6 + \&c.;$$

multiplying by dx, and integrating, we obtain,

$$\sin^{-1}x = x + \frac{1}{2}\frac{x^3}{3} + \frac{1}{2}\cdot\frac{3}{4}\frac{x^5}{5} + \frac{1}{2}\cdot\frac{3}{4}\cdot\frac{5}{6}\frac{x^7}{7} + \&c.,$$

the constant being 0 when the arc is estimated from the beginning of the first quadrant.

If we take the arc of 30°, the sine of which is equal to half the radius (Trig. Art. XIV), we shall have

$$\sin^{-1}30° = \frac{1}{2} + \frac{1}{2}\cdot\frac{1}{3}\cdot\frac{1}{2^3} + \frac{1}{2}\cdot\frac{3}{4}\cdot\frac{1}{5}\cdot\frac{1}{2^5} + \frac{1}{2}\cdot\frac{3}{4}\cdot\frac{5}{6}\cdot\frac{1}{7}\cdot\frac{1}{2^7} + \text{\&c.};$$

hence,

$$\pi = 6\sin^{-1}30° = 6\left(\frac{1}{2} + \frac{1.1.1}{2.3.2^3} + \frac{1.3.1.1}{2.4.5.2^5} + \frac{1.3.5.1.1}{2.4.6.7.2^7} + \text{\&c.}\right).$$

and by taking the first ten terms of the series, we find

$$\pi = 3.1415962,$$

which is true to the last decimal figure, which should be 5.

232. We will add a few more examples.

1. To integrate the expression $\dfrac{dx}{\sqrt{x - x^2}}$.

By making $\sqrt{x} = u$, we have

$$\frac{dx}{\sqrt{x - x^2}} = \frac{dx}{\sqrt{x}\sqrt{1 - x}} = \frac{2\,du}{\sqrt{1 - u^2}}.$$

But from the last series

$$\int\frac{2\,du}{\sqrt{1 - u^2}} = 2\left(u + \frac{1}{2}\frac{u^3}{3} + \frac{1}{2}\cdot\frac{3}{4}\frac{u^5}{5} + \frac{1}{2}\cdot\frac{3}{4}\cdot\frac{5}{6}\frac{u^7}{7} + \text{\&c.}\right) + C;$$

hence

$$\int\frac{dx}{\sqrt{x - x^2}} = 2\left(1 + \frac{1}{2}\frac{x}{3} + \frac{1}{2}\cdot\frac{3}{4}\frac{x^2}{5} + \frac{1}{2}\cdot\frac{3}{4}\cdot\frac{5}{6}\frac{x^3}{7} + \text{\&c.}\right)\sqrt{x} + C.$$

2. $$dx\sqrt{2ax - x^2} = (2a)^{\frac{1}{2}}x^{\frac{1}{2}}dx\left(1 - \frac{x}{2a}\right)^{\frac{1}{2}}.$$

But

$$\left(1-\frac{x}{2a}\right)^{\frac{1}{2}}=1-\frac{1}{2}\frac{x}{2a}-\frac{1}{2}\cdot\frac{1}{4}\frac{x^2}{4a^2}-\frac{1.1.3}{2.4.6}\frac{x^3}{8a^3}-\text{\&c.},$$

hence

$$\int dx\sqrt{2ax-x^2}=\left(\frac{2}{3}x^{\frac{3}{2}}-\frac{1}{2}\cdot\frac{2}{5}\frac{x^{\frac{5}{2}}}{2a}-\frac{1}{2}\cdot\frac{1}{4}\cdot\frac{2}{7}\frac{x^{\frac{7}{2}}}{4a^2}\right.$$

$$\left.-\frac{1}{2}\cdot\frac{1}{4}\cdot\frac{3}{6}\cdot\frac{2}{9}\frac{x^{\frac{9}{2}}}{8a^3}-\text{\&c.}\right)\sqrt{2a}+C:$$

and consequently

$$\int dx\sqrt{2ax-x^2}=\left(\frac{1}{3}-\frac{1}{2}\cdot\frac{1}{5}\frac{x}{2a}-\frac{1}{2}\cdot\frac{1}{4}\cdot\frac{1}{7}\frac{x^2}{4a^2}\right.$$

$$\left.-\frac{1}{2}\cdot\frac{1}{4}\cdot\frac{3}{6}\cdot\frac{1}{9}\frac{x^3}{8a^3}-\text{\&c.}\right)2x\sqrt{2ax}+C.$$

If the radius of a circle be represented by a, and the origin of co-ordinates be placed in the circumference, the equation will be (An. Geom. Bk. III, Prop. I, Sch. 3),

$$y^2=2ax-x^2;\quad\text{hence}\quad y=\sqrt{2ax-x^2},$$

and consequently (Art. 130)

$$dx\sqrt{2ax-x^2}=ydx$$

is the differential of a circular segment.

If we estimate the area from the origin, where $x=0$, we shall have $C=0$. If then we make $x=a$, the series will give the area of one quarter of the circle, if we make $x=2a$, of the semicircle.

3. $$\int\frac{dx}{\sqrt{1+x^2}}=x-\frac{1}{2}\frac{x^3}{3}+\frac{1.3}{2.4}\frac{x^5}{5}-\frac{1.3.5}{2.4.6}\frac{x^7}{7}+\text{\&c.}+C.$$

$$4. \int \frac{dx}{\sqrt{x^2-1}} = x - \frac{1}{2.2x^2} - \frac{1.3}{2.4.4x^4} - \frac{1.3.5}{2.4.6.6x^6} - \text{\&c.} + C.$$

Integration of Differential Binomials.

234. Differential binomials may be represented under the general form

$$x^{m-1}dx(a+bx^n)^{\frac{p}{q}},$$

in which, without affecting the generality of the expression, m and n may be regarded as entire numbers, and n as positive.

For, if m and n were fractional, and the binomial of the form

$$x^{\frac{1}{3}}dx(a+bx^{\frac{1}{2}})^{\frac{p}{q}}$$

make $x=z^6$, that is, make the exponent of z the least common multiple of the denominators of the exponents of x, and we shall then have

$$x^{\frac{1}{3}}dx(a+bx^{\frac{1}{2}})^{\frac{p}{q}} = 6z^7dz(a+bz^3)^{\frac{p}{q}},$$

in which the exponents of the variable are entire

If n were negative, we should have,

$$x^{m-1}dx(a+bx^{-n})^{\frac{p}{q}},$$

and by making $x=\frac{1}{z}$, we should obtain

$$-z^{-m-1}dz(a+bz^n)^{\frac{p}{q}},$$

the same form as before.

Furthermore, the binomial

$$x^{m-1}\,dx(ax^r + bx^n)^{\frac{p}{q}}$$

may be reduced to the form

$$x^{m+\frac{pr}{q}-1}\,dx(a + bx^{n-r})^{\frac{p}{q}},$$

by dividing the binomial within the parenthesis by x^r, and multiplying the factor without by $x^{\frac{pr}{q}}$.

235. Let us now determine the cases in which the binomial $x^{m-1}dx(a + bx^n)^{\frac{p}{q}}$ has an exact integral.

Make $a + bx^n = z^q$; we shall then have

$$x^n = \frac{z^q - a}{b}, \qquad (a + bx^n)^{\frac{p}{q}} = z^p, \qquad x^m = \left(\frac{z^q - a}{b}\right)^{\frac{m}{n}},$$

and by differentiating,

$$x^{m-1}\,dx = \frac{q}{nb}z^{q-1}\left(\frac{z^q - a}{b}\right)^{\frac{m}{n}-1} dz;$$

hence

$$x^{m-1}\,dx(a + bx^n)^{\frac{p}{q}} = \frac{q}{nb}z^{p+q-1}dz\left(\frac{z^q - a}{b}\right)^{\frac{m}{n}-1},$$

which will have an exact integral in algebraic terms when $\frac{m}{n}$ is a whole number and positive (Art. 216). If $\frac{m}{n}$ is negative see Art. 260.

Hence, *every differential binomial has an exact integral, when the exponent of the variable without the parenthesis augmented by unity, is exactly divisible by the exponent of the variable within.*

Thus, for example, the expression

$$x^5dx(a+bx^2)^{\frac{p}{q}}$$

has an exact integral. For, by comparing it with the general binomial, we find

$$m = 6, \ n = 2, \quad \text{and consequently,} \quad \frac{m}{n} = 3,$$

and the transformed binomial becomes

$$\frac{q}{2b} z^{p+q-1} dz \left(\frac{z^q - a}{b}\right)^2.$$

236. There is yet another case in which the binomial $x^{m-1} dx(a + bx^n)^{\frac{p}{q}}$ has an exact integral.

If we multiply and divide the quantity within the parenthesis by x^n, we have

$$x^{m-1} dx(a + bx^n)^{\frac{p}{q}} = x^{m-1} dx\left[(ax^{-n} + b)x^n\right]^{\frac{p}{q}}$$

$$= x^{m-1} dx(ax^{-n} + b)^{\frac{p}{q}} x^{\frac{np}{q}}$$

$$= x^{m+\frac{np}{q}-1} dx(ax^{-n} + b)^{\frac{p}{q}},$$

Now, if we add unity to the exponent of x without the parenthesis, and divide by $-n$, the quotient will be $-\left(\frac{m}{n} + \frac{p}{q}\right)$, and the expression will have an exact integral when this quotient is a whole number (Art. 235).

Hence, *every differential binomial has an exact integral, when the exponent of the variable without the parenthesis augmented by unity and divided by the exponent of the variable within the parenthesis, plus the exponent of the parenthesis, is an entire number.*

237. The integration of differential binomials is effected by resolving them into two parts, of which one at least has a known integral.

We have seen (Art. 28) that

$$d(uv) = udv + vdu,$$

whence, by integrating,

$$uv = \int udv + \int vdu,$$

and, consequently,

$$\int udv = uv - \int vdu.$$

Hence, if we have a differential of the form Xdx, in which the function X may be decomposed into two factors P and Q, of which one of them, Qdx, can be integrated, we shall have, by making $\int Qdx = v$ and $P = u$,

$$\int PQdx = Pv - \int vdP,$$

in which it is only required to integrate the term $\int vdP$.

238. To abridge the results, let us write p for $\frac{p}{q}$, in which case p will represent a fraction, and the differential binomial will take the form

$$x^{m-1}dx(a+bx^n)^p.$$

If now, we multiply by the two factors x^n and x^{-n}, the value will not be affected, and we obtain

$$x^{m-n}x^{n-1}dx(a+bx^n)^p.$$

Now, the factor $x^{n-1}dx(a+bx^n)^p$ is integrable, whatever be the value of p (Art. 217); and representing this factor by dv, we have

$$v = \frac{(a+bx^n)^{p+1}}{(p+1)nb}, \quad \text{and} \quad u = x^{m-n},$$

and, consequently,

$$\int x^{m-1}dx(a+bx^n)^p =$$

$$\frac{x^{m-n}(a+bx^n)^{p+1}}{(p+1)nb} - \frac{m-n}{(p+1)nb}\int x^{m-n-1}dx(a+bx^n)^{p+1}.$$

But, $\int x^{m-n-1}dx(a+bx^n)^{p+1} =$

$$\int x^{m-n-1}dx(a+bx^n)^p(a+bx^n) =$$

$$a\int x^{m-n-1}dx(a+bx^n)^p + b\int x^{m-1}dx(a+bx^n)^p;$$

substituting this last value in the preceding equation, and collecting the terms containing the integral

$$\int x^{m-1}dx(a+bx^n)^p,$$

we have

$$\left(1+\frac{m-n}{(p+1)n}\right)\int x^{m-1}dx(a+bx^n)^p =$$

$$\frac{x^{m-n}(a+bx^n)^{p+1} - a(m-n)\int x^{m-n-1}dx(a+bx^n)^p}{(p+1)nb};$$

whence,

formula (A.) . . . . $\int x^{m-1}dx(a+bx^n)^p =$

$$\frac{x^{m-n}(a+bx^n)^{p+1} - a(m-n)\int x^{m-n-1}dx(a+bx^n)^p}{b(pn+m)}.$$

This formula reduces the differential binomial

$\int x^{m-1}dx(a+bx^n)^p$ to that of $\int x^{m-n-1}dx(a+bx^n)^p$;

and by a similar process we should find

$\int x^{m-n-1}dx(a+bx^n)^p$ to depend on $\int x^{m-2n-1}dx(a+bx^n)^p$;

and consequently, each process diminishes the exponent of the variable without the parenthesis by the exponent of the variable within.

After the second integration, the factor $m-n$, of the second term, will become $m-2n$; and after the third, $m-3n$, &c. If m is a multiple of n, the factor $m-n$, $m-2n$, $m-3n$, &c., will finally become equal to 0, and then the differential into which it is multiplied will disappear, and the given differential will have an exact integral which corresponds with the result of Art. 235.

239. Let us now determine a formula for diminishing the exponent of the parenthesis.

We have

$$\int x^{m-1}dx(a+bx^n)^p = \int x^{m-1}dx(a+bx^n)^{p-1}(a+bx^n) =$$
$$a\int x^{m-1}dx(a+bx^n)^{p-1} + b\int x^{m+n-1}dx(a+bx^n)^{p-1}.$$

Applying formula (A) to the second term, by placing $m+n$ for m, and $p-1$ for p, we have

$$\int x^{m+n-1}dx(a+bx^n)^{p-1} =$$
$$\frac{x^m(a+bx^n)^p - am\int x^{m-1}dx(a+bx^n)^{p-1}}{b(pn+m)}.$$

Substituting this value in the last equation, we have

$$\text{formula (B)} \ldots\ldots\ldots\ldots\ldots\ldots \int x^{m-1}dx(a+bx^n)^p =$$
$$\frac{x^m(a+bx^n)^p + pna\int x^{m-1}dx(a+bx^n)^{p-1}}{pn+m},$$

which diminishes the exponent of the parenthesis by unity for each integration.

240. By means of formulas (A) and (B), we reduce

$$\int x^{m-1}dx(a+bx^n)^p \quad \text{to} \quad \int x^{m-rn-1}dx(a+bx^n)^{p-s};$$

rn being the greatest multiple of n which can be taken from $m-1$, and s the greatest whole number which can be subtracted from p.

For example, $\int x^7dx(a+bx^3)^{\frac{5}{2}}$ is reduced, by formula (A), to

$$\int x^4dx(a+bx^3)^{\frac{5}{2}}, \quad \text{and then to} \quad \int xdx(a+bx^3)^{\frac{5}{2}}:$$

and by formula (B) $\int x dx(a+bx^3)^{\frac{5}{2}}$, reduces to

$$\int x dx(a+bx^3)^{\frac{3}{2}}, \text{ and finally to } \int x dx(a+bx^3)^{\frac{1}{2}}.$$

241. It is evident that formulas (A) and (B) will only diminish the exponents $m-1$ and p, when m and p are positive. We will now determine two formulas for diminishing these exponents when they are negative.

We find from formula (A)

$$\int x^{m-n-1}dx(a+bx^n)^p =$$
$$\frac{x^{m-n}(a+bx^n)^{p+1}-b(m+np)\int x^{m-1}dx(a+bx^n)^p}{a(m-n)};$$

and placing for m, $-m+n$, we have

$$\text{formula (C)} \ldots\ldots\ldots\ldots \int x^{-m-1}dx(a+bx^n)^p =$$
$$\frac{x^{-m}(a+bx^n)^{p+1}+b(m-n-np)\int x^{-m+n-1}dx(a+bx^n)^p}{-am},$$

in which formula, it should be remembered that the negative sign has been attributed to the exponent m.

242. To find the formula for diminishing the exponent of the parenthesis when it is negative.

We find, from formula (B),

$$\int x^{m-1}dx(a+bx^n)^{p-1} =$$
$$-\frac{x^m(a+bx^n)^p-(m+np)\int x^{m-1}dx(a+bx^n)^p}{pna},$$

writing for p, $-p+1$, we have

$$\text{formula (D)} \ldots\ldots\ldots\ldots \int x^{m-1}dx(a+bx^n)^{-p} =$$
$$\frac{x^m(a+bx^n)^{-p+1}-(m+n-np)\int x^{m-1}dx(a+bx^n)^{-p+1}}{(p-1)na}.$$

This formula does not apply to the case in which $p=1$. Under this supposition, the second member becomes infinite, and the differential becomes that of a transcendental function.

243. It is sometimes convenient to leave the variable in both terms of the binomial. We shall therefore determine a particular formula for integrating the binomial

$$x^q(2ax-x^2)^{-\frac{1}{2}}dx=\frac{x^q\,dx}{\sqrt{2ax-x^2}}:$$

This binomial may be placed under the form

$$\int x^{q-\frac{1}{2}}dx(2a-x)^{-\frac{1}{2}},$$

and if we apply formula (A), after making

$$m=q+\frac{1}{2},\quad n=1,\quad p=-\frac{1}{2},\quad a=2a,\quad b=-1,$$

we shall have

$$\int x^{q-\frac{1}{2}}dx(2a-x)^{-\frac{1}{2}}=$$

$$-\frac{x^{q-\frac{1}{2}}(2a-x)^{\frac{1}{2}}}{q}+\frac{2a(q-\frac{1}{2})}{q}\int x^{q-\frac{3}{2}}dx(2a-x)^{-\frac{1}{2}};$$

and if we observe that

$$x^{q-\frac{1}{2}}=x^{q-1}x^{\frac{1}{2}}\quad x^{q-\frac{3}{2}}=x^{q-1}x^{-\frac{1}{2}},$$

and pass the fractional powers of x within the parentheses we shall have

formula (E) . . . . $\displaystyle\int\frac{x^q\,dx}{\sqrt{2ax-x^2}}=$

$$-\frac{x^{q-1}\sqrt{2ax-x^2}}{q}+\frac{(2q-1)a}{q}\int\frac{x^{q-1}dx}{\sqrt{2ax-x^2}},$$

which diminishes the exponent of the variable without the parenthesis by unity. If q is a positive and entire number, we shall have, after q reductions

$$\int \frac{dx}{\sqrt{2ax - x^2}} = \text{ver-sin}^{-1}\frac{x}{a}. \text{ (Art. 226).}$$

Integration of Rational Fractions.

244. Every rational fraction may be written under the form

$$\frac{Px^{n-1} + Qx^{n-2} \dots + Rx + S}{P'x^n + Q'x^{n-1} \dots + R'x + S'}dx,$$

in which the exponent of the highest power of the variable in the numerator, is less by unity than in the denominator. For, if the greatest exponent in the numerator was equal to or exceeded the greatest exponent in the denominator, the division might be made, giving one or more entire terms for a quotient and a remainder, in which the exponent of the leading letter would be less by at least unity, than the exponent of the leading letter in the divisor. The entire terms could then be integrated, and there would remain the fraction under the above form.

Place the denominator of the fraction equal to 0: that is, make

$$P'x^n + Q'x^{n-1} \dots\dots R'x + S' = 0,$$

and let us also suppose that we have found the n binomial factors into which it may be resolved (Alg. Art. 264). These factors will be of the form $x - a$, $x - b$, $x - c$ $x - d$, &c. Now there are three cases:

1st. When the roots of the equation are real and unequal.

2d. When they are real and equal.

3d. When there are imaginary factors.

We will consider these cases in succession.

1st. *When the roots are real and unequal.*

245. Let us take, as a first example, $\frac{adx}{x^2 - a^2}$.

By decomposing the denominator into its factors, we have

$$\frac{adx}{x^2 - a^2} = \frac{adx}{(x-a)(x+a)},$$

and we may make

$$\frac{adx}{(x-a)(x+a)} = \left(\frac{A}{x-a} + \frac{B}{x+a}\right)dx,$$

in which A and B are constants, whose values are yet to be determined. In order to determine these constants, let us reduce the terms of the second member of the equation to a common denominator; we shall then have

$$\frac{adx}{(x-a)(x+a)} = \frac{(Ax + Aa + Bx - Ba)dx}{(x-a)(x+a)}.$$

In comparing the two members of the equation, we find

$$a = Ax + Aa + Bx - Ba;$$

or, by arranging with reference to x,

$$(A+B)x + (A - B - 1)a = 0.$$

But, since this equation is true for all values of x, the

coefficients must be separately equal to 0 (Alg. Art. 208): hence

$$A+B=0, \quad \text{and} \quad (A-B-1)a=0,$$

which gives

$$A=\frac{1}{2}, \quad B=-\frac{1}{2},$$

Substituting these values for A and B, we obtain

$$\frac{adx}{x^2-a^2}=\frac{\frac{1}{2}dx}{x-a}-\frac{\frac{1}{2}dx}{x+a};$$

and integrating, we find (Art. 218)

$$\int\frac{adx}{x^2-a^2}=\frac{1}{2}\log(x-a)-\frac{1}{2}\log(x+a)+C,$$

and, consequently,

$$\int\frac{adx}{x^2-a^2}=\frac{1}{2}\log\left(\frac{x-a}{x+a}\right)+C=\log\left(\frac{x-a}{x+a}\right)^{\frac{1}{2}}+C.$$

246. Let us take, as a second example, $\dfrac{a^3+bx^2}{a^2x-x^3}dx.$ The factors of the denominator are x and a^2-x^2; but

$$a^2-x^2=(a+x)(a-x):$$

hence, the given fraction becomes

$$\frac{a^3+bx^2}{x(a-x)(a+x)}dx.$$

Let us now make

$$\frac{a^3+bx^2}{x(a-x)(a+x)}=\frac{A}{x}+\frac{B}{a-x}+\frac{C}{a+x},$$

reducing the terms of the second member to a common denominator, we have

$$\frac{a^3+bx^2}{x(a-x)(a+x)}=\frac{Aa^2-Ax^2+Bax+Bx^2+Cax-Cx^2}{x(a-x)(a+x)},$$

and, comparing the like powers of x (Alg. Art. 208),

$$B-A-C=b, \quad Ba+Ca=0, \quad Aa^2=a^3.$$

From these equations, we find

$$A=a, \quad B=\frac{a+b}{2}, \quad C=-\frac{a+b}{2},$$

and substituting these values, we obtain

$$\frac{a^3+bx^2}{a^2x-x^3}dx=a\frac{dx}{x}+\frac{a+b}{2(a-x)}dx-\frac{a+b}{2(a+x)}dx;$$

and integrating (Art. 218),

$$\int\frac{a^3+bx^2}{a^2x-x^3}dx=a\log x-\frac{a+b}{2}\log(a-x)$$

$$-\frac{a+b}{2}\log(a+x)+C$$

$$=a\log x-\frac{a+b}{2}[\log(a-x)+\log(a+x)]+C$$

$$=a\log x-\frac{a+b}{2}\log(a-x)(a+x)+C$$

$$=a\log x-\frac{a+b}{2}\log(a^2-x^2)+C$$

$$=a\log x-(a+b)\log\sqrt{a^2-x^2}+C.$$

247. Let us take, for a third example, $\dfrac{3x-5}{x^2-6x+8}dx.$

Resolving the denominator into the two binomial factors (Alg. Art. 142), $(x-2)$, $(x-4)$, we have

$$\frac{3x-5}{x^2-6x+8}=\frac{A}{x-2}+\frac{B}{x-4}, \quad \text{hence}$$

$$\frac{3x-5}{x^2-6x+8}=\frac{Ax-4A+Bx-2B}{x^2-6x+8};$$

and by comparing the coefficients of x, we have

$$-5=-4A-2B, \qquad 3=A+B,$$

which gives

$$B=\frac{7}{2}, \qquad A=-\frac{1}{2},$$

and substituting these values, we have

$$\int\frac{3x-5}{x^2-6x+8}dx=-\frac{1}{2}\int\frac{dx}{x-2}+\frac{7}{2}\int\frac{dx}{x-4}+C$$
$$=\frac{7}{2}\log(x-4)-\frac{1}{2}\log(x-2)+C.$$

248. Let us take, as a last example,

$$\frac{xdx}{x^2+4ax-b^2}.$$

Resolving the equation

$$x^2+4ax-b^2=0,$$

we find

$$x=-2a+\sqrt{4a^2+b^2}, \qquad x=-2a-\sqrt{4a^2+b^2},$$

and consequently, for the product of the factors,

$$(x+2a+\sqrt{4a^2+b^2})(x+2a-\sqrt{4a^2+b^2})=x^2+4ax-b^2.$$

To simplify the work, represent the roots by $-K$ and $-L$, and the factors will then be

$$x+K, \quad x+L,$$

and we shall have

$$\frac{x}{x^2+4ax-b^2}=\frac{A}{x+K}+\frac{B}{x+L}: \quad \text{hence}$$

$$\frac{x}{x^2+4ax-b^2}=\frac{Ax+AL+Bx+BK}{x^2+4ax-b^2},$$

whence,

$$AL+BK=0, \quad A+B=1,$$

and, consequently,

$$A=\frac{K}{K-L}, \qquad B=-\frac{L}{K-L}:$$

hence,

$$\int\frac{xdx}{x^2+4ax-b^2}=\frac{K}{K-L}\log(x+K)-\frac{L}{K-L}\log(x+L)+C.$$

249. In general, to integrate a rational fraction of the form

$$\frac{Px^{m-1}+Qx^{m-2}\ldots\ldots+Rx+S}{x^m\ +Q'x^{m-1}\ldots\ldots+R'x+S'}dx.$$

1st. *Resolve the fraction into* m *partial fractions, of which the numerators shall be constants, and the denominators factors of the denominator of the given fraction.*

2d. *Find the values of the numerators of the partial fractions, and multiply each by* dx.

3d. *Integrate each partial fraction separately, and the sum of the integrals thus found will be the integral sought.*

250. The method which has just been explained, will require some modification when any of the roots of the denominator are equal to each other. When the roots are unequal, the fraction may be placed under the form

$$\frac{Px^4 + Qx^3 + Rx^2 + Sx + T}{(x-a)(x-b)(x-c)(x-d)(x-e)}$$
$$= \frac{A}{x-a} + \frac{B}{x-b} + \frac{C}{x-c} + \frac{D}{x-d} + \frac{E}{x-e};$$

if several of these roots are equal, as for example, $a = b = c$, the last equation will become

$$\frac{Px^4 + Qx^3 + \&c.}{(x-a)^3(x-d)(x-e)} = \frac{A+B+C}{x-a} + \frac{D}{x-d} + \frac{E}{x-e}.$$

in which $A + B + C$ may be represented by a single constant A'.

Now, in reducing the second member of the equation to a common denominator with the first, and comparing the coefficients of the like powers of x, we shall have five equations of condition between three arbitrary constants, A', D, and E: hence, these equations will be incompatible with each other (Alg. Art. 103).

If, however, instead of adding the three partial fractions

$$\frac{A}{x-a}, \quad \frac{B}{x-a}, \quad \frac{C}{x-a},$$

which have the same denominator, we go through the

process of reducing them to one, their sum may be placed under the form

$$\frac{A' + B'x + C'x^2}{(x-a)^3},$$

or, by omitting the accents,

$$\frac{A + Bx + Cx^2}{(x-a)^3}.$$

251. Let us now make

$$x - a = z, \quad \text{and consequently,} \quad x = z + a;$$

we shall then have

$$\frac{A + Bx + Cx^2}{(x-a)^3} = \frac{A + Ba + Ca^2 + Bz + 2Caz + Cz^2}{z^3}$$

$$= \frac{A + Ba + Ca^2}{z^3} + \frac{B + 2Ca}{z^2} + \frac{C}{z};$$

substituting for z its value, and representing the numerators by single constants, we have

$$\frac{A + Bx + Cx^2}{(x-a)^3} = \frac{A'}{(x-a)^3} + \frac{B'}{(x-a)^2} + \frac{C'}{x-a};$$

the form under which the fraction may be written.

Since the same reasoning will apply to the case in which there are m equal factors, we conclude that

$$\frac{Px^{m-1} + Qx^{m-2} \ldots . + Rx + S}{(x-a)^m} =$$

$$\frac{A}{(x-a)^m} + \frac{A'}{(x-a)^{m-1}} + \frac{A''}{(x-a)^{m-2}} \ldots . + \frac{A''\ldots'}{x-a}.$$

252. In order, therefore, to integrate the fraction

$$\frac{Px^4 + Q'x^3 + \&c.}{(x-a)^3(x-d)(x-e)}dx,$$

place it equal to

$$\frac{A}{(x-a)^3}+\frac{A'}{(x-a)^2}+\frac{A''}{x-a}+\frac{D}{x-d}+\frac{E}{x-e};$$

then, reducing to a common denominator, and comparing the coefficients of the like powers of x, we find the values of the numerators of the partial fractions. Multiplying each by dx, and the given fraction may be written under the form

$$\frac{A}{(x-a)^3}dx+\frac{A'}{(x-a)^2}dx+\frac{A''}{(x-a)}dx+\frac{D}{x-d}dx+\frac{E}{x-e}dx.$$

The first two fractions may be integrated by the method of Art. 217, and the three last by logarithms. Hence, finally,

$$\int\frac{Px^4+Qx^3+Rx^2+Sx+T}{(x-a)^3(x-d)(x-e)}dx=-\frac{A}{2(x-a)^2}-\frac{A'}{x-a}$$
$$+A''\log(x-a)+D\log(x-d)+E\log(x-e)+C.$$

253. Let it be required to integrate the fraction

$$\frac{2ax}{(x+a)^2}dx.$$

We have

$$\frac{2ax}{(x+a)^2}=\frac{A}{(x+a)^2}+\frac{A'}{x+a};$$

reducing the fractions of the second member to a common denominator, and comparing the coefficients of x in the two members, we have

$$2a=A' \quad\text{and}\quad A+A'a=0:$$

hence,

$$A=-2a^2, \quad\text{and}\quad A'=2a;$$

and, consequently,

$$\frac{2axdx}{(x+a)^2} = -\frac{2a^2dx}{(x+a)^2} + \frac{2adx}{(x+a)};$$

hence, (Arts. 217 & 218),

$$\int \frac{2axdx}{(x^2+a)^2} = \frac{2a^2}{x+a} + 2a\log(x+a).$$

254. Let us find the integral of

$$\frac{x^2dx}{x^3 - ax^2 - a^2x + a^3}.$$

By placing the denominator equal to 0, we see that, by making $x = a$, the terms will destroy each other: hence, a is a root of the equation, and $x - a$ a factor. Dividing by $x - a$, the quotient is $x^2 - a^2$: hence, the fraction may be placed under the form

$$\frac{x^2dx}{(x^2-a^2)(x-a)} = \frac{x^2dx}{(x+a)(x-a)(x-a)}$$
$$= \frac{x^2dx}{(x-a)^2(x+a)}.$$

Let us now make

$$\frac{x^2}{(x-a)^2(x+a)} = \frac{A}{(x-a)^2} + \frac{A'}{(x-a)} + \frac{B}{x+a}.$$

Reducing the terms of the second member to a common denominator, we have

$$\frac{x^2}{(x-a)^2(x+a)} = \frac{A(x+a) + A'(x^2-a^2) + B(x-a)^2}{(x-a)^2(x+a)};$$

and developing, and comparing the coefficients of the like

powers of x, we obtain the equations

$$A' + B = 1, \qquad A - 2Ba = 0, \qquad Aa - A'a^2 + Ba^2 = 0.$$

Multiplying the first equation by a^2, and adding it to the third, we have

$$Aa + 2Ba^2 = a^2;$$

then multiplying the second by a, and adding it to the last, we have

$$a^2 = 2Aa, \quad \text{and consequently,} \quad A = \frac{1}{2}a;$$

substituting this value of A, we find

$$B = \frac{1}{4} \quad \text{and} \quad A' = \frac{3}{4}.$$

Substituting these values of A, A', and B, we have

$$\frac{x^2dx}{(x-a)^2(x+a)} = \frac{adx}{2(x-a)^2} + \frac{3dx}{4(x-a)} + \frac{dx}{4(x+a)},$$

and consequently,

$$\int \frac{x^2dx}{x^3 - ax^2 - a^2x + a^3} = -\frac{a}{2(x-a)} + \frac{3}{4}\log(x-a)$$
$$+ \frac{1}{4}\log(x+a) + C.$$

255. We can integrate, in a similar manner, when the denominator contains sets of equal roots. Let us take, as an example,

$$\frac{adx}{(x^2-1)^2} = \frac{adx}{(x-1)^2(x+1)^2}.$$

Make

$$\frac{a}{(x-1)^2(x+1)^2}=\frac{A}{(x-1)^2}+\frac{A'}{x-1}+\frac{B}{(x+1)^2}+\frac{B'}{x+1},$$

reducing the second member to a common denominator, we find the numerator equal to

$$A(x+1)^2+A'(x-1)(x+1)^2+B(x-1)^2+B'(x+1)(x-1)^2;$$

and comparing the coefficients with those of the numerator of the first member, we have the following equations:

$$\begin{aligned} A'+B'&=0, \\ A+A'+B-B'&=0, \\ 2A-A'-2B-B'&=0, \\ A-A'+B+B'&=a. \end{aligned}$$

Combining the first and third equations, we find $A=B$; and combining the second and fourth, gives $2A+2B=a$: hence, we have

$$A=B=\frac{a}{4}, \quad A'=-\frac{a}{4}, \quad B'=\frac{a}{4};$$

consequently, the given differential becomes

$$\frac{1}{4}a\left[\frac{dx}{(x-1)^2}+\frac{dx}{(x+1)^2}-\frac{dx}{x-1}+\frac{dx}{x+1}\right],$$

and by integrating,

$$\int\frac{adx}{(x^2-1)^2}=\frac{1}{4}a\left[-\frac{1}{x-1}-\frac{1}{x+1}-\log(x-1)+\log(x+1)\right]+C.$$

256. If an equation of the second degree has imaginary roots, the quantity under the radical sign will be essentially

negative (Alg. Art. 144), and the roots will be of the form

$$x = \mp a + b\sqrt{-1}, \qquad x = \mp a - b\sqrt{-1},$$

and the two binomial factors corresponding to the roots will be

$$(x \pm a - b\sqrt{-1})\,(x \pm a + b\sqrt{-1}) = x^2 \pm 2ax + a^2 + b^2.$$

Hence, for each set of imaginary roots which arise from placing the denominator of the fraction equal to 0, there will be a factor of the second degree of the form

$$x^2 \pm 2ax + a^2 + b^2.$$

257. If the imaginary roots are equal, we shall have,

$$a = 0, \quad x = +b\sqrt{-1}, \quad x = -b\sqrt{-1},$$

and the factor will become $x^2 + b^2$.

In the equation,

$$x^2 - 6cx + 10c^2 = 0,$$

the roots are,

$$x = 3c + c\sqrt{-1}, \quad x = 3c - c\sqrt{-1};$$

comparing these values of x with the general form, we have

$$a = -3c \qquad b = c,$$

and the given equation takes the form

$$x^2 - 6cx + 9c^2 + c^2 = 0.$$

Comparing the roots of the equation,

$$x^2 + 4x + 12 = 0,$$

with the values of x in the general form, we have

$$a = 2, \qquad b = \sqrt{8},$$

and the equation may be written under the form

$$x^2 + 4x + 4 + 8 = 0.$$

258. Let us first consider the case in which the denominator of the fraction to be integrated contains but one set of imaginary roots. The fraction will then be of the form,

$$\frac{P + Qx + Rx^2 + Sx^3 + \&c.}{(x-a)(x-b)\ldots\ldots(x-h)(x^2 + 2ax + a^2 + b^2)}dx,$$

which may be placed under the form

$$\frac{Adx}{x-a} + \frac{Bdx}{x-b}\ldots\ldots + \frac{Hdx}{x-h} + \frac{Mx + N}{x^2 + 2ax + a^2 + b^2}dx.$$

The first three fractions may be integrated by the methods already explained: it therefore only remains to integrate the last, which may be written under the form

$$\frac{Mx + N}{(x+a)^2 + b^2}dx.$$

If we make $x + a = z$, the expression becomes

$$\frac{Mz + N - Ma}{z^2 + b^2}dz,$$

and making $N - Ma = P$, it reduces to

$$\frac{Mz + P}{z^2 + b^2}dz,$$

which may be divided into the parts,

$$\frac{Mzdz}{z^2 + b^2} + \frac{Pdz}{z^2 + b^2},$$

which may be integrated separately.

To integrate the first term, we have

$$\int \frac{Mzdz}{z^2+b^2} = M\int \frac{zdz}{z^2+b^2} = \frac{M}{2}\int \frac{2zdz}{z^2+b^2},$$

in which the numerator, $2zdz$, is equal to the differential of the denominator: hence (Art. 218),

$$\int \frac{Mzdz}{z^2+b^2} = \frac{M}{2}\log(z^2+b^2);$$

or by substituting for z its value, $x+a$,

$$\int \frac{Mzdz}{z^2+b^2} = \frac{M}{2}\log[(x+a)^2+b^2]$$

$$= \frac{M}{2}\log(x^2+2ax+a^2+b^2)$$

$$= M\log\sqrt{x^2+2ax+a^2+b^2}.$$

Integrating the second term by Art. 224, gives

$$\int \frac{Pdz}{z^2+b^2} = \frac{P}{b}\text{tang}^{-1}\left(\frac{z}{b}\right),$$

or by substituting for z its value, $x+a$, and for P, $N-Ma$, we have

$$\int \frac{Pdz}{z^2+b^2} = \frac{N-Ma}{b}\text{tang}^{-1}\left(\frac{x+a}{b}\right);$$

and finally,

$$\int \frac{Mx+N}{x^2+2ax+a^2+b^2}dx =$$

$$M\log\sqrt{x^2+2ax+a^2+b^2} + \frac{N-Ma}{b}\text{tang}^{-1}\left(\frac{x+a}{b}\right).$$

259. Let us take, as an example, the fraction

$$\frac{c+fx}{x^3-1}dx,$$

in which, if $+1$ be substituted for x, the denominator will reduce to 0: hence, $x-1$ is a factor of the denominator. Dividing by this factor, the fraction may be put under the form

$$\frac{c+fx}{(x-1)(x^2+x+1)}dx,$$

in which x^2+x+1 is the product of the imaginary factors. Placing this product equal to 0, finding the roots of the equation, and comparing them with the general values in the form

$$x^2+2ax+a^2+b^2=0,$$

we find

$$a=\frac{1}{2} \qquad b=\sqrt{\frac{3}{4}}.$$

We may place the given fraction under the form

$$\frac{c+fx}{(x-1)(x^2+x+1)}=\frac{A}{x-1}+\frac{Mx+N}{x^2+x+1};$$

reducing the second member to a common denominator, and comparing the coefficients of x in the numerator with those of x in the numerator of the first member, we obtain

$$A=\frac{c+f}{3}, \quad M=-\frac{c+f}{3}, \quad N=\frac{f-2c}{3}.$$

Substituting these values of M and N, as also those of a and b, in the general formula of Art. 258, and recollecting that

$$\int\frac{Adx}{x-1}=\frac{c+f}{3}\int\frac{dx}{x-1}=\frac{c+f}{3}\log(x-1),$$

we find

$$\int \frac{c+fx}{x^3-1}dx = \frac{c+f}{3}\log(x-1) - \frac{c+f}{3}\log\sqrt{x^2+x+1}$$

$$+\frac{f-c}{\sqrt{3}}\text{tang}^{-1}\left[\frac{x+\frac{1}{2}}{\frac{1}{2}\sqrt{3}}\right]+C.$$

260. The equation which arises from placing the d nominator of the fraction equal to 0, may give severa. pairs of imaginary roots respectively equal to each other. In this case, the factor $x^2 \pm 2ax + a^2 + b^2$ will enter several times into the denominator, or will take the form

$$(x^2 + 2ax + a^2 + b^2)^p;$$

and hence, that part of the fraction which contains the pairs of equal and imaginary roots, must be placed under the form (Art. 251)

$$\frac{H + Kx}{(x^2 + 2ax + a^2 + b^2)^p} + \frac{H' + K'x}{(x^2 + 2ax + a^2 + b^2)^{p-1}}$$

$$+ \frac{H'' + K''x}{(x^2 + 2ax + a^2 + b^2)^{p-2}} \ldots\ldots + \frac{H^n + K^n x}{x^2 + 2ax + a^2 + b^2}.$$

Now, reducing to a common denominator, and comparing the coefficients, we find the values of the constants

$$H,\ K,\ H',\ K',\ H'',\ K'' \ldots\ldots H^n,\ K^n \ldots$$

after which, multiply each term by dx, and then integrate the terms separately.

Since all the terms are of the same general form, it will only be necessary to integrate the first term, which may be written under the form

$$\frac{H + Kx}{[(x+a)^2 + b^2]^p}dx;$$

which, if we make $x + a = z$, will reduce to

$$\frac{H - Ka + Kz}{(z^2 + b^2)^p} dz,$$

and making $M = H - Ka$, it will become

$$\frac{M + Kz}{(b^2 + z^2)^p} dz = \frac{Kzdz}{(b^2 + z^2)^p} dz + \frac{Mdz}{(b^2 + z^2)^p}.$$

The first term of the second member may be placed under the form

$$K \int (b^2 + z^2)^{-p} zdz,$$

and integrating by the formula of Art. 217, we have

$$\int \frac{Kzdz}{(b^2 + z^2)^p} = \frac{1}{2} \frac{K}{(1 - p)} \frac{1}{(b^2 + z^2)^{p-1}} + C.$$

It then only remains to integrate the second term

$$\frac{Mdz}{(b^2 + z^2)^p} = M \int (b^2 + z^2)^{-p} dz.$$

By comparing the second member of this equation with formula (D), Art. 242, we see that it will become identical with the first member of that formula, by supposing

$$m = 1, \quad a = b^2, \quad b = 1, \quad \text{and} \quad n = 2;$$

and hence, by means of that formula, the exponent $-p$ may be successively diminished by unity until it becomes -1, when the integration of the term will depend on that of

$$\frac{dz}{b^2 + z^2}.$$

But we have already found (Art. 224),

$$\int \frac{dz^2}{b^2 + z^2} = \frac{1}{b} \text{tang}^{-1}\left(\frac{z}{b}\right);$$

and hence the fraction may be considered as entirely integrated.

261. It follows, from the preceding discussion, that the integration of all rational fractions depends on the following forms:

$$\text{1st.} \qquad \int x^m dx = \frac{x^{m+1}}{m+1}.$$

$$\text{2d.} \qquad \int \frac{dx}{a \pm x} = \pm \log(a \pm x).$$

$$\text{3d.} \qquad \int \frac{dx}{a^2 + x^2} = \frac{1}{a} \operatorname{tang}^{-1}\left(\frac{x}{a}\right).$$

Integration of Irrational Fractions.

262. The method of integrating rational fractions having been explained, we may consider an irrational fraction as admitting of integration when it is reduced to a rational form.

263. Every irrational fraction in which the radical quantities are monomials, may be reduced to a rational form.

Let us take, as an example,

$$\frac{\sqrt{x} - \frac{1}{3}a}{\sqrt[3]{x} - \sqrt{x}} dx, \qquad \text{or} \qquad \frac{x^{\frac{1}{2}} - \frac{1}{3}a}{x^{\frac{1}{3}} - x^{\frac{1}{2}}}.$$

Having found the least common multiple of the indices of the roots, (which indices are the denominators of the fractional exponents,) substitute for x a new variable, z, with this common multiple for an exponent, and the fraction will then become rational in terms of z.

In the example given, the least common multiple is 6; hence we have

$$x = z^6 \quad \text{and} \quad \sqrt{x} = z^3, \quad \sqrt[3]{x} = z^2, \quad dx = 6z^5dz;$$

and substituting these values, we obtain

$$\frac{\sqrt{x} - \frac{1}{3}a}{\sqrt[3]{x} - \sqrt{x}} = \frac{z^3 - \frac{1}{3}a}{z^2 - z^3} 6z^5dz = \frac{6z^6 - 2az^3}{1 - z} dz;$$

an expression which may be integrated by rational fractions; after which we may substitute for z its value, $\sqrt[6]{x}$.

264. If the quantity under the radical sign is a polynomial, the fraction cannot, in general, be reduced to a rational form. We can, however, reduce to a rational form every expression of the form

$$X(\sqrt{A + Bx \pm Cx^2})dx,$$

in which X is supposed to be a rational function of x.

If we write a denominator 1, and then multiply the numerator and denominator by $\sqrt{A + Bx \pm Cx^2}$, the expression will take the form

$$\frac{X'dx}{\sqrt{A + Bx \pm Cx^2}};$$

in which X' is a rational function of x: hence the two forms are essentially the same.

If now, we can find rational values for $\sqrt{A + Bx \pm Cx^2}$ and for dx, in terms of a new variable, the expression will take a rational form.

There are two cases to be considered: 1st., when the coefficient of x^2 is positive; and, 2d, when it is negative

Let us consider them separately. First, make

$$\sqrt{A+Bx+Cx^2}=\sqrt{C}\sqrt{\frac{A}{C}+\frac{B}{C}x+x^2}$$

$$=\sqrt{C}\sqrt{a+bx+x^2},$$

in which $a=\frac{A}{C}$, $b=\frac{B}{C}$.

In order to find rational values for dx and $\sqrt{a+bx+x^2}$, place

$$\sqrt{a+bx+x^2}=x+z, \quad (1)$$

from which, by squaring both members, we find

$$a+bx=2xz+z^2, \quad (2)$$

and hence,

$$x=\frac{z^2-a}{b-2z}; \quad (3)$$

and substituting this value in equation (1),

$$\sqrt{a+bx+x^2}=\frac{z^2-a}{b-2z}+z;$$

and by reducing to the same denominator,

$$\sqrt{a+bx+x^2}=-\frac{z^2-bz+a}{b-2z}. \quad (4)$$

Let us now find the value of dx in terms of z. For this purpose we will differentiate equation (2), we then find

$$bdx=2xdz+2zdx+2zdz;$$

whence we have

$$(b-2z)dx=2(x+z)dz;$$

and by subtracting equations (1) and (4), and substituting for $x+z$ the value thus found, we have

$$(b-2z)dx = -\frac{2(z^2-bz+a)}{b-2z}dz,$$

and
$$dx = -\frac{2(z^2-bz+a)}{(b-2z)^2}dz. \quad (5)$$

265. Let us take, as an example,

$$\frac{dx}{x\sqrt{A+Bx+Cx^2}},$$

which may be written under the form

$$\frac{dx}{\sqrt{C}\times x\sqrt{a+bx+x^2}};$$

and substituting the values of $\sqrt{a+bx+x^2}$ and dx, from equations (4) and (5), we have

$$\frac{dx}{\sqrt{a+bx+x^2}} = \frac{2dz}{b-2z};$$

and multiplying the denominator by the value of x, in equation (3),

$$\frac{dx}{x\sqrt{a+bx+x^2}} = \frac{2dz}{z^2-a};$$

and then by $\sqrt{C}$, we have

$$\frac{dx}{\sqrt{C}\times x\sqrt{a+bx+x^2}}, \text{ or } \frac{dx}{x\sqrt{A+Bx+Cx^2}} = \frac{2dz}{(z^2-a)\sqrt{C}},$$

which is a rational form, and may be integrated by the methods already explained.

266. Let us take, as a second example,

$$\frac{dx}{\sqrt{h + c^2x^2}},$$

which may be placed under the form

$$\frac{dx}{c\sqrt{\frac{h}{c^2} + x^2}};$$

and comparing this with the form of Art. 264, gives

$$c = \sqrt{C}, \quad b = 0, \quad \frac{h}{c^2} = a.$$

Hence,

$$\int \frac{dx}{\sqrt{h + c^2x^2}} = \frac{1}{c}\int \frac{dx}{\sqrt{a + x^2}}.$$

Having placed

$$\sqrt{a + x^2} = z + x,$$

we found, Art. 264, equations (5) and (4),

$$dx = -\frac{z^2 + a}{2z^2}dz, \quad \sqrt{a + x^2} = \frac{z^2 + a}{2z};$$

hence

$$\int \frac{dx}{\sqrt{a + x^2}} = \int -\frac{dz}{z} = -\log z.$$

Substituting for z its value, and multiplying by $\frac{1}{c}$, we have

$$\int \frac{dx}{\sqrt{h + c^2x^2}} = -\frac{1}{c}\log[\sqrt{a + x^2} - x] + C,$$

and substituting for a its value, $\frac{h}{c^2}$, we have

$$\int \frac{dx}{\sqrt{h + c^2x^2}} = -\frac{1}{c}\log\left[\frac{1}{c}(\sqrt{h + c^2x^2} - cx)\right] + C$$

$$= -\frac{1}{c}\log\frac{1}{c} - \frac{1}{c}\log(\sqrt{h + c^2x^2} - cx) + C.$$

But since the difference of the squares of the two terms within the parenthesis is equal to h, it follows that if h be divided by the difference of the terms, the quotient will be their sum (Alg. Art. 59). But the division may be effected by subtracting their logarithms. Let us, then, add to, and subtract from, the second member of the equation, $\frac{1}{c}\log h$. We shall then have,

$$\int \frac{dx}{\sqrt{h+c^2x^2}} = -\frac{1}{c}\log\frac{1}{c} - \frac{1}{c}\log h + \frac{1}{c}\log h - \frac{1}{c}\log(\sqrt{h+c^2x^2} - cx) + C;$$

or by representing the three constants $-\frac{1}{c}\log\frac{1}{c} - \frac{1}{c}\log h$, and C, by a single letter C, we have

$$\int \frac{dx}{\sqrt{h+c^2x^2}} = \frac{1}{c}\log(\sqrt{h + c^2x^2} + cx) + C.$$

267. Let us take, as a third example,

$$dx\sqrt{m^2 + x^2}.$$

Comparing this with the general form, we find

$$a = m^2 \quad \text{and} \quad b = 0;$$

hence (Art. 264),

$$\sqrt{m^2 + x^2} = \frac{z^2 + m^2}{2z} \quad \text{and} \quad dx = -\frac{(z^2 + m^2)}{2z^2}dz;$$

and consequently,

$$dx\sqrt{m^2+x^2} = -\frac{(z^2+m^2)^2}{4z^3}dz,$$

which is rational in z; and, having found the integral in z, substitute the value of z in terms of x.

268. Let us now consider the case in which the coefficient of x^2 is negative. We have

$$\sqrt{A+Bx-Cx^2} = \sqrt{C}\sqrt{\frac{A}{C}+\frac{B}{C}-x^2}$$

$$= \sqrt{C}\,\sqrt{a+bx-x^2}.$$

If now, we make as before,

$$\sqrt{a+bx-x^2} = x+z,$$

and square both members, the second powers of x in each member will not cancel, as before; and therefore, x cannot be expressed rationally in terms of z. We must, therefore, place the value of the radical under another form. We will remark, in the first place, that the binomial $a+bx-x^2$, may be decomposed into two rational factors of the first degree, with respect to x. For, if we make

$$x^2-bx-a=0,$$

and designate the roots of the equation by α and α', we have (Alg. Art. 142)

$$(x^2-bx-a) = (x-\alpha)(x-\alpha'),$$

and consequently, by changing the signs,

$$(a+bx-x^2) = -(x-\alpha)(x-\alpha') = (x-\alpha)(\alpha'-x),$$

and placing the second member under the radical, we may make

$$\sqrt{(x-\alpha)(\alpha'-x)}=(x-\alpha)z; \quad (1)$$

squaring both members

$$(x-\alpha)(\alpha'-x)=(x-\alpha)^2z^2,$$

and by suppressing the common factor $x-\alpha$,

$$\alpha'-x=(x-\alpha)z^2, \quad (2)$$

whence,

$$x=\frac{\alpha'+\alpha z^2}{1+z^2},$$

and
$$x-\alpha=\frac{\alpha'+\alpha z^2}{1+z^2}-\alpha;$$

or by reducing,

$$x-\alpha=\frac{\alpha'-\alpha}{1+z^2}; \quad (3)$$

which, being substituted in the second member of equation (1), gives

$$\sqrt{(x-\alpha)(\alpha'-x)}=\frac{\alpha'-\alpha}{1+z^2}z; \quad (4)$$

and by differentiating equation (3), we obtain

$$dx=-\frac{2(\alpha'-\alpha)}{(1+z^2)^2}zdz. \quad (5)$$

269. To apply this method to a particular example of the form

$$\frac{dx}{\sqrt{a+bx-x^2}},$$

substitute the values of $\sqrt{a+bx-x^2}$ and dx, found in equations (4) and (5): we find

$$\frac{dx}{\sqrt{a+bx-x^2}} = -\frac{2(\alpha'-\alpha)z}{(1+z^2)^2 z\frac{(\alpha'-\alpha)}{1+z^2}}dz = -\frac{2dz}{1+z^2}$$

hence

$$\int\frac{dx}{\sqrt{a+bx-x^2}} = -2\,\text{tang}^{-1}z + C;$$

or, by substituting for z its value from equation (1),

$$\int\frac{dx}{\sqrt{a+bx-x^2}} = C - 2\,\text{tang}^{-1}\left(\frac{\sqrt{(x-\alpha)(\alpha'-x)}}{(x-\alpha)}\right)$$

$$= C - 2\,\text{tang}^{-1}\sqrt{\frac{\alpha'-x}{x-\alpha}}.$$

270. If, in the last formula, we make

$$a = 1 \quad\text{and}\quad b = 0,$$

the trinomial under the radical will become $1-x^2$, and the roots of the equation $x^2-1=0$ are

$$\alpha = -1 \quad\text{and}\quad \alpha' = 1.$$

Substituting these values, and the general formula becomes

$$\int\frac{dx}{\sqrt{1-x^2}} = C - 2\,\text{tang}^{-1}\sqrt{\frac{1-x}{1+x}};$$

and if we suppose the integral to be 0 when $x=0$, we shall have

$$0 = C - 2\,\text{tang}^{-1}(1)$$

$$= C - 2(45^\circ) \quad \text{(Trig. Art. VIII)}$$

$$= C - 90^\circ: \quad \text{hence} \quad C = \frac{\pi}{2}.$$

Substituting this value, and we have

$$\int \frac{dx}{\sqrt{1-x^2}} = \frac{\pi}{2} - 2\,\text{tang}^{-1}\sqrt{\frac{1-x}{1+x}}.$$

271. We have already seen (Art. 219) that

$$\int \frac{dx}{\sqrt{1-x^2}} = \sin^{-1} x;$$

and hence,

$$\frac{\pi}{2} - 2\,\text{tang}^{-1}\sqrt{\frac{1-x}{1+x}}$$

should also represent the arc of which x is the sine

To prove this, we have (Trig. Art. XXV)

$$\text{tang}\,2A = \frac{2\,\text{tang}\,A}{1-\text{tang}^2 A}.$$

Substituting for $\text{tang}\,A$, $\sqrt{\frac{1-x}{1+x}}$, and reducing, we have

$$\text{tang}\,2A = \frac{\sqrt{1-x^2}}{x};$$

that is, twice the arc whose tangent is $\sqrt{\frac{1-x}{1+x}}$ is equal to the arc whose tangent is $\frac{\sqrt{1-x^2}}{x}$.

But the arc whose tangent is $\frac{\sqrt{1-x^2}}{x}$, is the complement of the arc whose tangent is $\frac{x}{\sqrt{1-x^2}}$, (Trig. Art. XVIII); and this arc has x for its sine. Hence, either member of the equation

$$\int \frac{dx}{\sqrt{1-x^2}} = \frac{\pi}{2} - 2 \operatorname{tang}^{-1} \sqrt{\frac{1-x}{1+x}},$$

represents the arc whose sign is x.

272. Let us take, as a last example, the differential

$$dx \sqrt{2ax - x^2}.$$

In comparing this with the general form, we find (Art 268)

$$\alpha = 0 \quad \text{and} \quad \alpha' = 2a;$$

and Art. 268, equations (4) and (5), give

$$\sqrt{x(2a-x)} = \frac{2az}{1+z^2}, \quad dx = -\frac{4az}{(1+z^2)^2} dz.$$

Substituting these values, we have

$$dx \sqrt{2ax - x^2} = -\frac{8a^2z^2dz}{(1+z^2)^3};$$

which may be integrated by the method of rational fractions.

Rectification of Plane Curves.

273. The rectification of a curve is the expression for its length. When this expression can be found in a finite number of algebraic terms, the curve is said to be *rectifiable*, and its length may be represented by a straight line.

274. The differential of the arc of a curve, referred to rectangular co-ordinates, is (Art. 128)

$$dz = \sqrt{dx^2 + dy^2}.$$

Hence, if it be required to rectify a curve, given by its equation,

1st. *Differentiate the equation of the curve.*

2d. *Combine the differential equation thus found with the given equation, and find the value of* dx² *or* dy² *in terms of the other variable and its differential.*

3d. *Substitute the value thus found in the differential of the arc, which will then involve but one variable and its differential. Then, by integrating, we shall find an expression for the length of the arc, estimated from a given point, in terms of one of the co-ordinates.*

275. Let us take, as a first example, the common parabola, of which the equation is

$$y^2 = 2px.$$

Differentiating, and dividing by 2, we have

$$ydy = pdx,$$

and consequently,

$$dx^2 = \frac{y^2}{p^2}dy^2;$$

substituting this value in the differential of the arc, we have

$$dz = \sqrt{dy^2 + \frac{y^2}{p^2}dy^2}$$

$$= \frac{1}{p}dy\sqrt{p^2 + y^2};$$

which, being integrated by formula (B) Art. 239, gives by supposing $m = 1$, $a = p^2$, $b = 1$, $n = 2$, $p = \frac{1}{2}$,

$$\int dy\sqrt{p^2+y^2} = \frac{y\sqrt{p^2+y^2}}{2} + \frac{p^2}{2}\int\frac{dy}{\sqrt{p^2+y^2}};$$

and integrating the second term by the formula of Art. 266, we have, after making $h=p^2$, $c^2=1$,

$$\int\frac{dy}{\sqrt{p^2+y^2}} = \log(\sqrt{p^2+y^2}+y);$$

and consequently,

$$z=\frac{1}{p}\int dy\sqrt{p^2+y^2}=\frac{y\sqrt{p^2+y^2}}{2p}+\frac{p}{2}\log(\sqrt{p^2+y^2}+y)+C$$

If we estimate the arc from the vertex of the parabola, we shall have

$$y=0 \quad \text{for} \quad z=0: \quad \text{hence}$$

$$0=\frac{p}{2}\log p+C \qquad \text{or} \quad C=-\frac{p}{2}\log p;$$

and consequently,

$$z=\frac{y\sqrt{p^2+y^2}}{2p}+\frac{p}{2}\log\left(\frac{\sqrt{p^2+y^2}+y}{p}\right);$$

and hence, the value of the arc, for a given ordinate y, can only be found approximatively.

276. The curves represented by the equation

$$y^n=px^m,$$

are called *parabolas*. This equation may be placed under the form

$$y=p^{\frac{1}{n}}x^{\frac{m}{n}};$$

or by placing $p^{\frac{1}{n}}=p'$, and $\frac{m}{n}=n'$, we have

$$y=p'x^{n'};$$

or finally, by omitting the accents, the form becomes

$$y = px^n.$$

By differentiating, we have

$$dy = npx^{n-1}dx,$$

and by substituting this value of dy in the differential of the arc, we have

$$z = \int (1 + n^2p^2x^{2n-2})^{\frac{1}{2}} dx.$$

The integral of this expression will be expressed in a finite number of algebraic terms when $\frac{1}{2n-2}$ is a whole number and positive (Art. 235). If we designate such whole and positive number by i, we have for the condition of an exact integral in algebraic terms,

$$\frac{1}{2n-2} = i, \text{ or } n = \frac{2i+1}{2i};$$

and substituting for n, we have

$$y = px^{\frac{2i+1}{2i}}; \text{ or } y^{2i} = p^{2i}x^{2i+1},$$

which expresses the relation between x and y when the length of the arc can be found in finite algebraic terms. There is yet another case in which the integral will be expressed in finite and algebraic terms, viz. when $\frac{1}{2n-2} + \frac{1}{2}$ is a positive whole number (Art. 236 and 235.)

277. If we make $i = 1$, we have $n = \frac{3}{2}$, and

$$y^2 = p^2x^3,$$

which is the equation of the cubic parabola.

Under this supposition, the arc becomes (Art. 217)

$$z = \int (1 + n^2p^2x^{2n-2})^{\frac{1}{2}}dx = \frac{8}{27p^2}(1 + \frac{9}{4}p^2x)^{\frac{3}{2}} + C;$$

and hence, the cubic parabola is rectifiable (Art. 273).

If we estimate the arc from the vertex of the curve, we have $x = 0$, for $z = 0$: hence

$$0 = \frac{8}{27p^2} + C, \quad \text{or} \quad C = -\frac{8}{27p^2};$$

and consequently,

$$z = \frac{8}{27p^2}\left[\left(1 + \frac{9}{4}p^2x\right)^{\frac{3}{2}} - 1\right].$$

278. If the origin of co-ordinates is at the centre of the circle, the equation of the circumference is

$$R^2 = x^2 + y^2,$$

and the value of the arc,

$$z = R\int \frac{dx}{\sqrt{R^2 - x^2}}.$$

If the origin be placed on the curve

$$y^2 = 2Rx - x^2,$$

and

$$z = R\int \frac{dx}{\sqrt{2Rx - x^2}},$$

both of which expressions may be integrated by series, and the length of the arc found approximatively.

279. It remains to rectify the transcendental curves. The differential equation of the cycloid is (Art. 182)

$$dx = \frac{ydy}{\sqrt{2ry - y^2}},$$

which gives

$$dx^2 = \frac{y^2dy^2}{2ry - y^2}.$$

Substituting this value of dx^2 in the differential of the arc, we obtain

$$dz = \sqrt{dy^2 + \frac{y^2dy^2}{2ry - y^2}} = dy\sqrt{\frac{2ry}{2ry - y^2}}$$

$$= dy\sqrt{\frac{2r}{2r - y}} = (2r)^{\frac{1}{2}}(2r - y)^{-\frac{1}{2}}dy.$$

But (Art. 217)

$$\int (2r - y)^{-\frac{1}{2}}dy = -2(2r - y)^{\frac{1}{2}} + C;$$

and hence,

$$z = -(2r)^{\frac{1}{2}}2\sqrt{2r - y} + C = -2\sqrt{2r(2r - y)} + C.$$

If now, we estimate the arc z from B, the point at which $y = 2r$, we shall have, for $z = 0$, $y = 2r$; hence

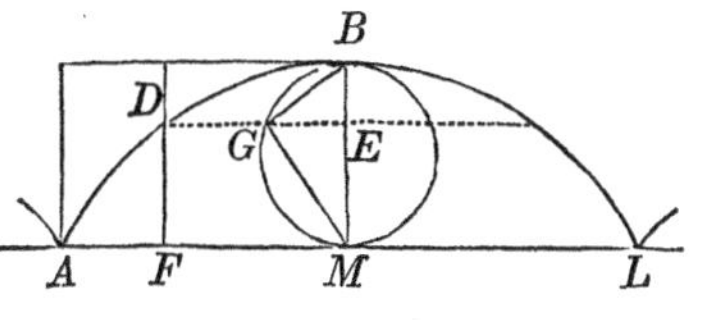

$$0 = 0 + C, \quad \text{or} \quad C = 0,$$

and consequently, the true integral will be

$$z = -2\sqrt{2r(2r - y)};$$

the second member being negative, since the arc is a decreasing function of the ordinate y (Art. 31).

If now, we suppose y to decrease until it becomes equal to any ordinate, as $DF = ME$, DB will be represented by z, or by $2\sqrt{2r(2r - y)}$, and $BE = 2r - y$.

But $\overline{BG}^2 = BM \times BE$: hence

$$BG = \sqrt{2r(2r - y)},$$

and consequently

$$BD = 2BG;$$

or *the arc of the cycloid, estimated from the vertex of the axis, is equal to twice the corresponding chord of the generating circle: hence, the arc* BDA *is equal to twice the diameter* BM; *and the curve* ADBL *is equal to four times the diameter of the generating circle.*

280. The differential of the arc of a spiral, referred to polar co-ordinates, is (Art. 202)

$$dz = \sqrt{du^2 + u^2dt^2}.$$

Taking the general equation of the spirals

$$u = at^n,$$

we have $$du^2 = n^2a^2t^{2n-2}dt^2;$$

and substituting for du^2 and u^2 their values, we obtain

$$dz = at^{n-1}dt\sqrt{n^2 + t^2}.$$

If we make $n = 1$, we have the spiral of Archimedes, (Art. 191), and the equation becomes

$$dz = adt\sqrt{1 + t^2};$$

which is of the same form as that of the arc of the common parabola (Art. 275).

281. In the logarithmic spiral, we have $t = \log u$, and the differential of the arc becomes

$$dz = du\sqrt{2} + C;$$

and if we estimate the arc from the pole,

$$z = u\sqrt{2}.$$

Consequently, the length of the arc estimated from the pole to any point of the curve, is equal to the diagonal of a square described on the radius-vector, although the number of revolutions of the radius-vector between these two points is infinite.

Of the Quadrature of Curves.

282. The quadrature of a curve is the expression of its area. When this expression can be found in finite algebraic terms, the curve is said to be *quadrable*, and may be represented by an equivalent square.

283. If s represents the area of the segment of a curve, and x and y the co-ordinates of any point, we have seen (Art. 130), that

$$ds = ydx.$$

To apply this formula to a given curve:

1st. *Find from the equation of the curve the value of* y *in terms of* x, *or the value of* dx *in terms of* y, *which values will be expressed under the forms*

$$y = f(x), \quad \text{or} \quad dx = f(y)dy.$$

2d. *Substitute the value of* y, *or the value of* dx, *in the differential of the area: we shall have*

$$ds = f(x)dx, \quad \text{or} \quad ds = f(y)dy:$$

the integral of the first form will give the area of the curve in terms of the abscissa, and the integral of the second will give the area in terms of the ordinate.

284. Let us take, as a first example, the family of parabolas of which the equation is

$$y^n = px^m:$$

we shall then have

$$y = p^{\frac{1}{n}} x^{\frac{m}{n}},$$

and

$$\int F(x)\,dx = \int p^{\frac{1}{n}} x^{\frac{m}{n}}\,dx = \frac{np^{\frac{1}{n}}}{m+n} x^{\frac{m+n}{n}} = \frac{n}{m+n} xy + C;$$

by substituting y for its value, $p^{\frac{1}{n}} x^{\frac{m}{n}}$.

If, instead of substituting the value of y in the differential of the area

$$ydx,$$

we find the value of dx from the equation

$$y^n = px^m,$$

we have

$$dx = \frac{n}{m} \frac{y^{\frac{n}{m}-1}}{p^{\frac{1}{m}}} dy,$$

and consequently,

$$\int ydx = \frac{n}{m} \int \frac{y^{\frac{n}{m}}}{p^{\frac{1}{m}}} dy = \frac{n}{m+n} \frac{y^{\frac{n}{m}+1}}{p^{\frac{1}{m}}} = \frac{n}{m+n} xy:$$

by substituting x for its value, $\frac{y^{\frac{n}{m}}}{p^{\frac{1}{m}}}$, which is the same result as before found.

Hence, *the area of any portion of a parabola is equal to the rectangle described on the abscissa and ordinate multiplied by the ratio* $\frac{n}{m+n}$. The parabolas are therefore quadrable.

In the common parabola, $n=2$, $m=1$, and we have

$$\int f(x)\,dx=\frac{2}{3}\,xy,$$

that is, *the area of a segment is equal to two thirds of the area of the rectangle described on the abscissa and ordinate.*

285. If, in the equation

$$y^n=px^m,$$

we make $n=1$, and $m=1$, it will represent a straight line passing through the origin of co-ordinates, and we shall have

$$\int f(x)\,dx=\frac{1}{2}xy,$$

which proves that *the area of a triangle is equal to half the product of the base and perpendicular.*

286. It is frequently necessary to find the integral or function, between certain limits of the variable on which it depends.

A particular notation has been adopted to express such integrals.

Resuming the equation of the common parabola

$$y^2=2px,$$

and substituting in the equation ydx the value of $dx=\frac{ydy}{p}$, we have

$$\int ydx=\frac{1}{p}\int y^2dy=\frac{y^3}{3p}+C;$$

or, if the area be estimated from the vertex A, we have $C=0$, and

$$\int y dx = \frac{y^3}{3p}.$$

If now, we wish the area to terminate at any ordinate $PM=b$, we shall then take the integral between the limits of $y=0$ and $y=b$; and, to express that in the differential equation, we write

$$\frac{1}{p}\int_0^b y^2 dy = \frac{b^3}{3p};$$

which is read, integral of y^2dy between the limits $y=0$ and $y=b$.

If we wish the area between the ordinates $MP=b$, $M'P'=c$, we must integrate between the limits $y=b$, $y=c$. We first integrate between 0 and each limit, viz.:

$$AMP = \frac{1}{p}\int_0^b y^2 dy = \frac{b^3}{3p},$$

$$AMM'P' = \frac{1}{p}\int_0^c y^2 dy = \frac{c^3}{3p}:$$

we then have

$$PMM'P = AMM'P' - AMP = \frac{1}{p}\int_b^c y^2 dy$$

$$= \frac{c^3}{3p} - \frac{b^3}{3p} = \frac{1}{3p}(c^3 - b^3).$$

287. Let us now determine the area of any portion of the space included between the asymptotes and curve of an hyperbola.

The equation of the hyperbola referred to its asymptotes (An. Geom. Bk. VI, Prop. IX,) is

$$xy = M.$$

In the differential of the area of a curve ydx, x and y are estimated in parallels to co-ordinate axes, at right angles to each other.

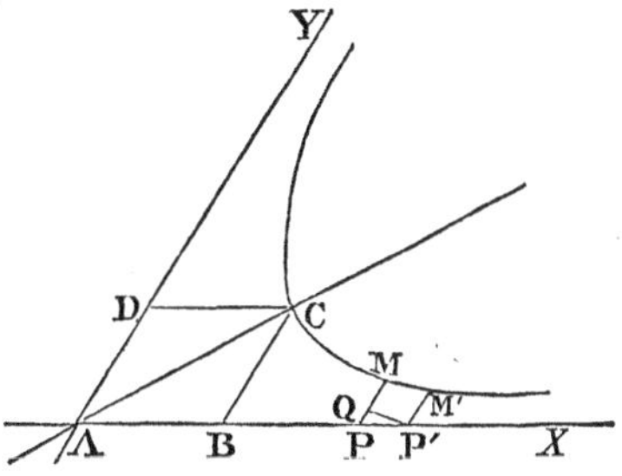

The differential of the area $BCMP$, referred to the oblique axes AX, AY, is the parallelogram $PMM'P'$, of which $PM = y$ and $PP' = dx$.

If we designate the angle $YAX = MPX$ by β, we shall have

$$\text{area } PMM'P = ydx\sin\beta;$$

and substituting for y its value $\frac{M}{x}$, and representing the area $BCMP$ by s, we have

$$ds = M\sin\beta\frac{dx}{x},$$

and
$$s = M\sin\beta\int\frac{dx}{x} = M\sin\beta\log x + C.$$

If AC is the semi-transverse axis of the hyperbola, and we make $AB = 1$, and estimate the area s from BC, we shall have, for $x = 1$, $s = 0$, and consequently $C = 0$; and the true integral will be

$$s = M\sin\beta\log x.$$

But, since $ABCD$ is a rhombus, and $M = AB \times BC$ (An. Geom. Bk. VI, Prop. IX, Sch. 2), and since $AB = 1$, we have $M = 1$, and consequently,

$$s = \sin\beta \log x.$$

Now, since s, which represents the space $BCMP$ for any abscissa x, is equal to the Naperian logarithm of x multiplied by the constant $\sin\beta$, s may be regarded as the logarithm of x taken in a system of which $\sin\beta$ is the modulus (Alg. Art. 268). Therefore, *any hyperbolic space* BCMP *is the logarithm of the corresponding abscissa* AP, *taken in the system whose modulus is the sine of the angle included between the asymptotes.*

If we would make the spaces the Naperian logarithms of the corresponding abscissas, we make $\sin\beta = 1$, which corresponds to the equilateral hyperbola. If we would make the spaces the common logarithms of the abscissas, make $\sin\beta = 0.43429945$, (Alg. Art. 272).

288. The equation of the circle, when the origin of co-ordinates is placed on the circumference, is

$$y^2 = 2rx - x^2, \quad \text{or} \quad y = \sqrt{2rx - x^2},$$

and hence, the differential of the area is

$$dx\sqrt{2rx - x^2};$$

and this will become, by making $x = r - u$,

$$-\int du(r^2 - u^2)^{\frac{1}{2}}.$$

If we integrate this expression by formula (B, Art. 239,

we have

$$-\int du(r^2-u^2)^{\frac{1}{2}} = -\frac{1}{2}u(r^2-u^2)^{\frac{1}{2}} - \frac{1}{2}r^2\int du(r^2-u^2)^{-\frac{1}{2}}$$

$$= -\frac{1}{2}u\sqrt{r^2-u^2} + \frac{1}{2}r^2 \int\frac{-du}{\sqrt{r^2-u^2}}.$$

But we have (Art. 253)

$$\int\frac{-du}{\sqrt{r^2-u^2}} = \cos^{-1}\left(\frac{u}{r}\right);$$

and placing for u its value

$$\int dx\sqrt{2rx-x^2} =$$

$$-\frac{1}{2}(r-x)\sqrt{2rx-x^2} + \frac{1}{2}r^2\cos^{-1}\left(\frac{r-x}{r}\right) + C;$$

and taking this integral between the limits $x=0$ and $x=2r$, we shall have the area of a semicircle.

For $x=0$, the area which is expressed in the first member becomes 0, the first term in the second member becomes 0, and the second term also becomes 0, since the arc whose cosine is 1, is 0: hence the constant $C=0$.

If we now make $x=2r$, the term

$$\frac{1}{2}(r-x)\sqrt{2rx-x^2}$$

reduces to 0, and the second term to

$$\frac{1}{2}r^2\cos^{-1}(-1) = \frac{1}{2}r^2\pi \quad \text{(Trig. Art. XIV)},$$

and consequently, the entire area is equal to $r^2\pi$, which corresponds with a known result (Geom. Bk. V, Prop. XII, Cor. 2).

The equation of the ellipse, the origin of co-ordinates being at the vertex of the transverse axis (An. Geom. Bk. IV, Prop. I. Sch. 8), gives

$$y = \frac{B}{A}\sqrt{2Ax - x^2},$$

and consequently, the area of the semi-ellipse will be equal to

$$\int y dx = \frac{B}{A}\int dx \sqrt{2Ax - x^2}.$$

Integrating, as in the last example, between the limits $x = 0$, and $x = 2A$, and multiplying by 2, we find $AB\pi$ for the entire area. This corresponds with a known result (An. Geom. Bk. IV, Prop. XIII).

289. The differential equation of the cycloid (Art. 183) is

$$dx = \frac{y dy}{\sqrt{2ry - y^2}},$$

whence

$$\int y dx = \int \frac{y^2 dy}{\sqrt{2ry - y^2}},$$

and applying formula E, (Art. 243) twice, it will reduce to

$$\int \frac{dy}{\sqrt{2ry - y^2}}; \text{ and (Art. 226)}$$

$$\int \frac{dy}{\sqrt{2ry - y^2}} = \text{ver-sin}^{-1}\left(\frac{y}{r}\right).$$

But we may determine the area of the cycloid in a more simple manner by introducing the exterior segment $AFKH$,

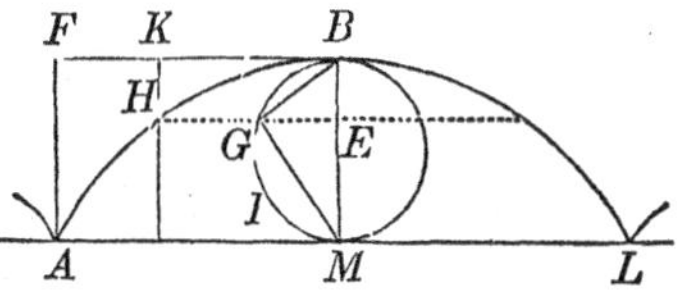

Regarding FB as a line of abscissas, and designating any ordinate as KH, by $z = 2r - y$, we shall have

$$d(AFKH) = zdx.$$

But

$$zdx = \frac{(2r - y)ydy}{\sqrt{2ry - y^2}} = dy\sqrt{2ry - y^2},$$

whence

$$AFKH = \int dy\sqrt{2ry - y^2} + C$$

But this integral expresses the area of the segment of a circle, of which the abscissa is y and radius r (Art. 288): that is, of the segment $MIGE$. If now, we estimate the area of the segment from M, where $y = 0$, and the area $AFKH$ from AF, in which case the area $AFKH = 0$ for $y = 0$, we shall have

$$AFKH = MIGE;$$

and taking the integral between the limits $y = 0$ and $y = 2r$, we have

$$AFB = \text{semicircle } MIGB,$$

and consequently,

$$\text{area } AHBM = AFBM - MIGB.$$

But the base of the rectangle $AFBM$ is equal to the semicircumference of the generating circle, and the altitude is equal to the diameter, hence its area is equal to four times the area of the semicircle $MIGB$; therefore,

$$\text{area } AHBM = 3\,MIGB,$$

and consequently, the *area* AHBL *is equal to three times the area of the generating circle.*

290. It now remains to determine the area of the spirals. If we represent by s the area described by the radius-vector, we have (Art. 203)

$$ds = \frac{u^2dt}{2};$$

and placing for u its value at^n (Art. 189)

$$ds = \frac{a^2t^{2n}dt}{2} \quad \text{and} \quad s = \frac{a^2t^{2n+1}}{4n+2} + C,$$

and if n is positive $C = 0$, since the area is 0 when $t = 0$. After one revolution of the radius-vector, $t = 2\pi$, and we have

$$s = \frac{a^2(2\pi)^{2n+1}}{4n+2},$$

which is the area included within the first spire.

291. In the spiral of Archimedes (Art. 192)

$$a = \frac{1}{2\pi} \quad \text{and} \quad n = 1;$$

hence, for this spiral we have

$$s = \frac{t^3}{24\pi^2},$$

which becomes $\frac{\pi}{3}$, after one revolution of the radius-vector; the unit of the number $\frac{\pi}{3}$ being a square whose side is unity. Hence, the area included by the first spire, is equal to one third the area of the circle whose radius is equal to the radius-vector after the first revolution.

In the second revolution, the radius-vector describes a

second time the area described in the first revolution; and in any revolution, it will pass over, or redescribe, all the area before generated. Hence, to find the area at the end of the mth revolution, we must integrate between the limits

$$t = (m-1)2\pi \quad \text{and} \quad t = m.2\pi,$$

which gives

$$\frac{m^3 - (m-1)^3}{3}\pi.$$

If it be required to find the area between any two spires, as between the mth and the $(m+1)$th, we have for the whole area to the $(m+1)$th spire equal to

$$\frac{(m+1)^3 - m^3}{3}\pi;$$

and subtracting the area to the mth spire, gives

$$\frac{(m+1)^3 - 2m^3 + (m-1)^3}{3}\pi = 2m\pi,$$

for the area between the mth and $(m+1)$th spires.

If we make $m = 1$, we shall have the area between the first and second spires equal to 2π: hence, *the area between the* m*th and* (m + 1)*th spires, is equal to* m *times the area between the first and second.*

292. In the hyperbolic spiral $n = -1$, and we have

$$ds = \frac{a^2t^{-2}}{2}dt \quad \text{and} \quad s = -\frac{a^2}{2t}.$$

The area s will be infinite when $t = 0$, but we can find the area included between any two radius-vectors b and c by integrating between the limits $t = b$, $t = c$, which will give

$$s = \frac{a^2}{2}\left(\frac{1}{b} - \frac{1}{c}\right).$$

293. In the logarithmic spiral $t = \log u$: hence, $dt = \frac{du}{u}$,

$$\frac{u^2 dt}{2} = \frac{udu}{2};$$

hence,
$$s = \int \frac{udu}{2} = \frac{u^2}{4} + C;$$

and by considering the area $s = 0$ when $u = 0$, we have $C = 0$ and

$$s = \frac{u^2}{4}.$$

Determination of the Area of Surfaces of Revolution.

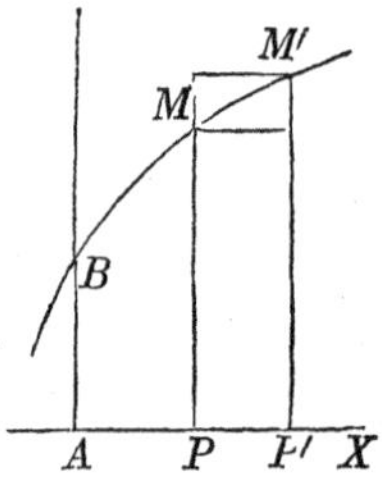

294. If any curve BMM', be revolved about an axis AX, it will describe a surface of revolution, and every plane passing through the axis AX will intersect the surface in a meridian curve. It is required to find the differential of this surface. For this purpose, make $AP = x$, $PM = y$, and $PP' = h$: we shall then have

$$PM = f(x) = y,$$

$$P'M' = f(x+h) = y + \frac{dy}{dx}h + \frac{d^2y}{dx^2}\frac{h^2}{1.2} + \text{\&c.}$$

In the revolution of the curve BMM', the extremities M and M' of the ordinates MP, $M'P'$, will describe the circumferences of two circles, and the chord MM' will describe the curved surface of the frustum of a cone. The surface of this frustum is equal to (Geom: Bk. VIII, Prop. IV.)

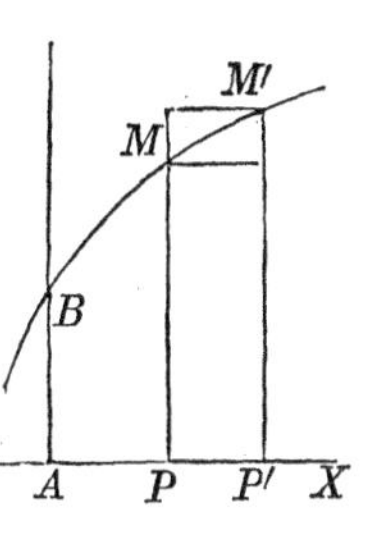

$$\frac{(circ. MP + circ. M'P')}{2} \times chord\ MM' : \qquad \text{that is, to}$$

$$\frac{(2\pi MP + 2\pi M'P')}{2} \times chord\ MM' = \pi (MP + M'P') \times chord\ MM';$$

and by substituting for MP, $M'P'$ their values, the expression for the area becomes

$$\pi \left(2y + \frac{dy}{dx}h + \frac{d^2y}{dx^2}\frac{h^2}{1.2} + \&c.\right) chord\ MM'.$$

If now we pass to the limit, by making $h = 0$, the chord MM' will become equal to the arc MM' (Art. 128), and the surface of the frustum of the cone will coincide with that of the surface described by the curve at the point M. If we represent the surface by s and the arc of the curve by z, we have, after passing to the limit,

$$ds = 2\pi y dz,$$

and by substituting for dz its value (Art. 128), we have

$$ds = 2\pi y \sqrt{dx^2 + dy^2}:$$

whence, *the differential of a surface of revolution is equal to the circumference of a circle perpendicular to the axis, into the differential of the arc of the meridian curve.*

Remark. It should be observed that X is the axis about which the curve is revolved. If it were revolved about the axis Y, it would be necessary to change x into y and y into x.

295. If a right angled triangle CAB be revolved about the perpendicular CA, the hypothenuse CB will describe the surface of a right cone. If we represent the base BA of the triangle by b, the altitude CA by h, and suppose the origin of co-ordinates at the vertex of the angle C, we shall have

$$x : y :: h : b: \quad \text{hence}$$

$$y = \frac{b}{h}x \quad \text{and} \quad dy = \frac{b}{h}dx.$$

Substituting these values of y and dy, in the general formula, we have

$$\int 2\pi y \sqrt{dx^2 + dy^2} = \int 2\pi \frac{bx}{h^2} dx \sqrt{h^2+b^2} = \pi \frac{bx^2}{h^2} \sqrt{h^2+b^2} + C,$$

and integrating between the limits $x = 0$ and $x = h$, we obtain

$$\text{surface of the cone } = \pi b \sqrt{h^2 + b^2} = 2\pi b \times \frac{CB}{2}$$

$$= circ. AB \times \frac{CB}{2}.$$

296. If a rectangle $ABCD$ be revolved around the side AD, we can readily find the surface of the right cylinder which will be described by the side BC.

Let us suppose the axis $AD = h$, and $AB = b$: the equation of the line DC will be $y = b$: hence, $dy = 0$. Substituting these values in the general expression of the differential of the surface, we have

$$\int 2\pi y \sqrt{dx^2 + dy^2} = \int 2\pi bdx = 2\pi bx + C;$$

and taking the integral between the limits $x=0$, $x=h$, we have

$$\text{surface} = 2\pi bh = circ. AB \times AD.$$

297. To find the surface of a sphere, let us take the equation of the meridian curve, referred to the centre as an origin: it is

$$x^2 + y^2 = R^2,$$

and by differentiating, we have

$$xdx + ydy = 0;$$

hence

$$dy = -\frac{xdx}{y} \quad \text{and} \quad dy^2 = \frac{x^2dx^2}{y^2}.$$

Substituting for dy its value, in the differential of the surface

$$ds = 2\pi y\sqrt{dx^2 + dy^2},$$

we obtain

$$s = \int 2\pi y \sqrt{dx^2 + \frac{x^2}{y^2}dx^2} = \int 2\pi R dx = 2\pi Rx + C.$$

If we estimate the surface from the plane passing through the centre, and perpendicular to the axis of X, we shall have

$$s = 0 \quad \text{for} \quad x = 0, \quad \text{and consequently} \quad C = 0.$$

Now, to find the entire surface of the sphere, we must integrate between the limits $x = +R$ and $x = -R$, and then take the sum of the integrals without reference to their algebraic signs, for these signs only indicate the position of the parts of the surface with respect to the plane passing through the centre of the sphere.

Integrating between the limits

$$x = 0 \quad \text{and} \quad x = +R,$$

we find

$$s = 2\pi R^2;$$

and integrating between the limits $x = 0$ and $x = -R$, there results

$$s = -2\pi R^2;$$

hence,

$$\text{surface} = 4\pi R^2 = 2\pi R \times 2R;$$

that is, equal to four great circles, or equal to the curved surface of the circumscribing cylinder.

298. The two equal integrals

$$s = 2\pi R^2 \quad \text{and} \quad s = -2\pi R^2$$

indicate that the surface is symmetrical with respect to the plane passing through the centre.

299. To find the surface of the paraboloid of revolution, take the equation of the meridian curve

$$y^2 = 2px,$$

which being differentiated, gives

$$dx = \frac{ydy}{p} \quad \text{and} \quad dx^2 = \frac{y^2dy^2}{p^2}.$$

Substituting this value of dx in the differential of the sur face, it reduces to

$$2\pi y\sqrt{\left(\frac{y^2+p^2}{p^2}\right)dy^2} = \frac{2\pi}{p}ydy\sqrt{y^2+p^2}$$

But we have found (Art. 217)

$$\int \frac{2\pi}{p} y dy \sqrt{y^2 + p^2} = \frac{2\pi}{3p}(y^2 + p^2)^{\frac{3}{2}} + C:$$

hence,

$$s = \frac{2\pi}{3p}(y^2 + p^2)^{\frac{3}{2}} + C,$$

and if we estimate the surface from the vertex at which point $y = 0$, we shall have,

$$0 = \frac{2\pi p^2}{3} + C, \quad \text{whence,} \quad C = -\frac{2\pi p^2}{3},$$

and integrating between the limits

$$y = 0, \quad y = b,$$

we have

$$s = \frac{2\pi}{3p}[(b^2 + p^2)^{\frac{3}{2}} - p^3]$$

300. To find the surface of an ellipsoid described by revolving an ellipse about the transverse axis.

The equation of the meridian curve is

$$A^2 y^2 + B^2 x^2 = A^2 B^2,$$

whence

$$dy = -\frac{B^2 x dx}{A^2 \; y} = -\frac{B}{A}\frac{x dx}{\sqrt{A^2 - x^2}}:$$

substituting the square of this value in the differential of the surface and for y its value

$$\frac{B}{A}\sqrt{A^2 - x^2}$$

we have

$$ds = 2\pi \frac{B}{A^2} dx \sqrt{A^4 - (A^2 - B^2)x^2},$$

and $$s = 2\pi \frac{B}{A^2}\sqrt{A^2 - B^2}\int dx \sqrt{\frac{A^4}{A^2 - B^2} - x^2};$$

and if we represent the part without the sign of the integral by D, and make

$$\frac{A^4}{A^2 - B^2} = R^2,$$

we shall have

$$s = D\int dx\sqrt{R^2 - x^2}.$$

But the integral of $dx\sqrt{R^2 - x^2}$ is a circular segment of which the abscissa is x, the radius of the circle being R. If, then, we estimate the surface of the ellipsoid from the plane passing through the centre, and also estimate the area of the circular segment from the same point, any portion of the surface of the ellipsoid will be equal to the corresponding portion of the circle multiplied by the constant D. Hence, if we integrate the expression

$$s = \int dx\sqrt{R^2 - x^2}$$

between the limits $x = 0$ and $x = A$, and designate by D' the corresponding portion of the circle whose radius is R, we shall have

$$\frac{1}{2}\text{ surface ellipsoid} = D \times D';$$

hence, $$\text{surface ellipsoid} = 2D \times D'.$$

301. To find the surface described by the revolution of the cycloid about its base.

The differential equation of the cycloid is

$$dx = \frac{ydy}{\sqrt{2ry - y^2}}.$$

Substituting this value of dx in the differential equation of the surface, it becomes

$$ds = \frac{2\pi\sqrt{2r}\,y^{\frac{3}{2}}\,dy}{\sqrt{2ry - y^2}}.$$

Applying formula (E), Art. 243, we have

$$s = 2\pi\sqrt{2r}\left[-\frac{2}{3}y^{\frac{1}{2}}\sqrt{2ry - y^2} + \frac{4}{3}r\int\frac{y^{\frac{1}{2}}\,dy}{\sqrt{2ry - y^2}}\right].$$

But,

$$\int\frac{y^{\frac{1}{2}}\,dy}{\sqrt{2ry - y^2}} = \int\frac{dy}{\sqrt{2r - y}} = \int dy\,(2r - y)^{-\frac{1}{2}} = -2(2r - y)^{\frac{1}{2}};$$

hence,

$$s = 2\pi\sqrt{2r}\left[-\frac{2}{3}y^{\frac{1}{2}}\sqrt{2ry - y^2} - \frac{8}{3}r(2r - y)^{\frac{1}{2}}\right] + C.$$

If we estimate the surface from the plane passing through the centre, we have $C = 0$, since at this point $s = 0$ and $y = 2r$. If we then integrate between the limits $y = 2r$ and $y = 0$, we have

$$s = \frac{1}{2}\text{ surface } = -\frac{32}{3}\pi r^2;\text{ hence,}$$

$$s = \text{surface } = -\frac{64}{3}\pi r^2,$$

that is, the surface described by the cycloid, when it is revolved around the base, is equal to 64 thirds of the generating circle.

The minus sign should appear before the integral, since the surface is a decreasing function of the variable y (Art. 31).

Of the Cubature of Solids of Revolution.

302. The cubature of a solid is the expression of its volume or content.

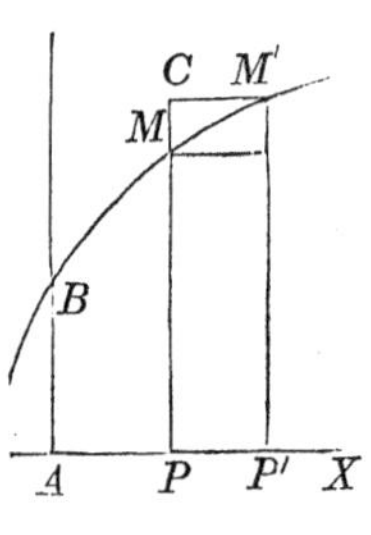

303. Let u represent the volume or solidity generated by the area $ABMP$, when revolved around the axis AX. If we make $AP=x$, $PP'=h$, we have $M'P'=F(x+h)$. Now, the solid generated by the area $ABMM'P'$, will exceed the solid described by $ABMP$, by the solid described by the area $PMM'P'$.

The solid described by the area $ABMP$ is a function of x, and the solid described by the area $ABMM'P'$ is a similar function of $(x+h)$. If we designate this last by u', we have

$$u'=u+\frac{du}{dx}h+\frac{d^2u}{dx^2}\frac{h^2}{1.2}+\frac{d^3u}{dx^3}\frac{h^3}{1.2.3}+\&c.;$$

hence, the solid described by $PMM'P'$ is

$$u'-u=\frac{du}{dx}h+\frac{d^2u}{dx^2}\frac{h^2}{1.2}+\frac{d^3u}{dx^3}\frac{h^3}{1.2.3}+\&c.$$

Let us now compare the cylinder described by the rectangle $P'M$ with that described by the rectangle $P'C$. The equation of the curve gives

$$MP=y=F(x) \quad M'P'=F(x+h);$$

hence, since $PP'=h$,

cylinder described by $P'M=\pi[F(x)]^2h$,

cylinder described by $P'C=\pi[F(x+h)]^2h$;

and the ratio of the cylinders is

$$\frac{[F(x+h)]^2}{[F(x)]^2};$$

the limit of which, when $h=0$, is unity.

But the solid described by the area $PMM'P'$ is less than one of the cylinders and greater than the other, hence, the limit of the ratio, when compared with either of them, is unity. Hence,

$$\frac{\frac{du}{dx}h+\frac{d^2u}{dx^2}\frac{h^2}{1.2}+\&c.}{\pi[F(x)]^2h}=\frac{\frac{du}{dx}+\frac{d^2u}{dx^2}\frac{h}{1.2}+\&c.}{\pi[F(x)]^2};$$

the limit of which, when $h=0$, is

$$\frac{\frac{du}{dx}}{\pi[F(x)]^2}=1,$$

whence

$$\frac{du}{dx}=\pi[F(x)]^2=\pi y^2,$$

and finally

$$du=\pi y^2dx;$$

the differential of the solidity πy^2dx being a cylinder whose base is πy^2 and altitude dx.

304. *Remark.* The differential of a solid, generated by revolving a curve around the axis of Y, is

$$\pi x^2dy.$$

305. Let it be required to find the solidity of a right cylinder with a circular base, the radius of the base being

r and the altitude h. We have for the differential of the solidity

$$\pi y^2 dx,$$

and since $y = r$, it becomes

$$\pi r^2 dx;$$

and taking the integral between the limits $x = 0$ and $x = h$ we have

$$\pi r^2 h,$$

which expresses the solidity.

306. To find the solidity of a right cone with a circular base, let us represent the altitude by h and the radius of the base by r, and let us also suppose the origin of co-ordinates at the vertex. We shall then have

$$y = \frac{r}{h}x \quad \text{and} \quad y^2 = \frac{r^2}{h^2}x^2;$$

and substituting, the differential of the solidity becomes

$$\frac{r^2}{h^2}\pi x^2 dx,$$

and by taking the integral between the limits $x = 0$ and $x = h$, we obtain

$$\frac{1}{3}r^2 \pi h = \pi r^2 \times \frac{h}{3};$$

that is, the area of the base into one third of the altitude.

307. Let it be required to find the solidity of a prolate spheroid, (An: Geom: Bk. IX, Art. 33).

The equation of a meridian section is

$$A^2y^2 + B^2x^2 = A^2B^2,$$

which gives

$$y^2 = \frac{B^2}{A^2}(A^2 - x^2);$$

hence the differential of the solidity is

$$du = \pi\frac{B^2}{A^2}(A^2 - x^2)dx,$$

and by integrating

$$u = \pi\frac{B^2}{A^2}\left(A^2x - \frac{x^3}{3}\right) + C$$

$$= \frac{\pi B^2}{3A^2}(3A^2x - x^3) + C.$$

If we estimate the solidity from the plane passing through the centre, we have for $x = 0$, $u = 0$, and consequently $C = 0$; and taking the integral between the limits $x = 0$ and $x = A$, we have

$$\frac{1}{2}\text{ solidity} = \frac{2}{3}\pi B^2 \times A;$$

and consequently

$$\text{solidity} = \frac{2}{3}\pi B^2 \times 2A.$$

But πB^2 expresses the area of a circle described on the conjugate axis, and $2A$ is the transverse axis: hence, *the solidity is equal to two-thirds of the circumscribing cylinder.*

308. If an ellipse be revolved around the conjugate axis, it will describe an oblate spheroid, and the differential of the solidity would be

$$du = \pi x^2 dy:$$

and substituting for x^2, and integrating, we should find

$$\text{solidity} = \frac{2}{3}\pi A^2 \times 2B:$$

that is, two-thirds of the circumscribing cylinder.

309. If we compare the two solids together, we find

$$\text{oblate spheroid} : \text{prolate spheroid} :: A : B.$$

310. If we make $A = B$, we obtain the solidity of the sphere, which is equal to two-thirds of the circumscribing cylinder, or equal to

$$\frac{4}{3}\pi R^3 = \frac{1}{6}\pi D^3.$$

311. Let it be required to find the solidity of a paraboloid. The equation of a meridian section is

$$y^2 = 2px,$$

and hence the differential of the solidity is

$$du = 2\pi px dx; \quad \text{hence}$$

$$u = \pi px^2 + C;$$

and estimating the solidity from the vertex, and taking the integral between the limits $x = 0$ and $x = h$, and designating by b the ordinate corresponding to the abscissa $x = h$, we have

$$u = \pi ph^2 = \pi b^2 \times \frac{h}{2};$$

that is, equal to half the cylinder having an equal base and altitude.

312. Let it be required, as a last example, to determine

the solidity of the solid generated by the revolution of the cycloid about its base.

The differential equation of the cycloid is

$$dx = \frac{ydy}{\sqrt{2ry - y^2}};$$

hence we have

$$du = \frac{\pi y^3 dy}{\sqrt{2ry - y^2}};$$

which, being integrated by formula (E) Art. 243, and then by Art. 226, we find the solidity equal to five-eighths of the circumscribing cylinder.

Of Double Integrals.

313. Let us, in the first place, consider a solid limited by the three co-ordinate planes, and by a curved surface which is intersected by the co-ordinate planes in the curves CB, BD, DC.

Through any point of the surface, as M, pass two planes HQF and EPG respectively parallel to the co-ordinate planes ZX, YZ, and intersecting the surface in the curves HMF and EMG. The co-ordinates of the point M are

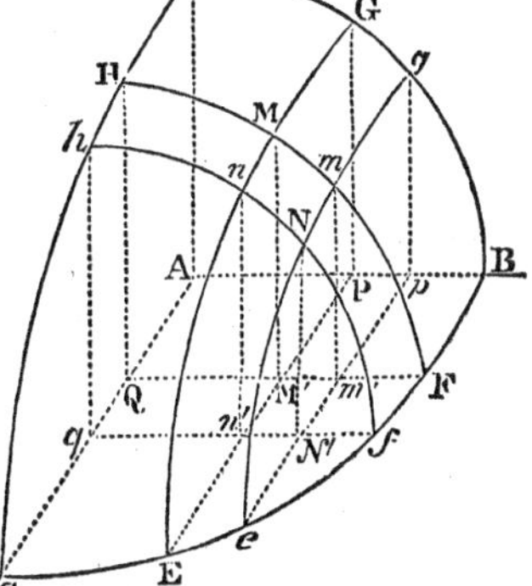

$AP = x$, $PM' = y$, $MM' = z$.

It is now evident that the solid whose base on the co-ordinate plane YX is the rectangle $AQM'P$, may be extended indefinitely in the direction of the axis of X without changing the value of y, or indefinitely in the direction of Y without changing x. Hence, x and y may be regarded as independent variables.

If, for example, we suppose y to remain constant, and x to receive an increment $Pp=h$, the solid whose base is the rectangle $AQM'P$, will be increased by the solid whose base is the rectangle $M'm'pP$; and if we suppose x to remain constant, and y to receive an increment $Qq=k$, the first solid will be increased by the solid whose base is the rectangle $Qqn'M'$.

But if we suppose x and y to receive their increments at the same time, the new solid will still be bounded by the parallel planes epg, hqf, and will differ from the primitive solid not only by the two solids before named, but also by the solid whose base is the rectangle $n'M'm'N'$. This last solid is the increment of the solid whose base is the rectangle $M'Ppm'$, when we suppose y to vary; or the increment of the solid whose base is the rectangle $Qqn'M'$, when we suppose x to vary.

Let us represent by u the solid whose base is the rectangle $AQM'P$; u will then be a function of x and y, and the difference between the values of the increments of u, under the supposition that x and y vary separately; and under the supposition that they vary together, will be equal to the solid whose base is the rectangle $n'M'm'N'$. By taking this difference (Art. 83) we have

$$\text{solid}\, n'N'm'M' \ldots M = \frac{d^2u}{dxdy}hk + \frac{1}{2}\frac{d^3u}{dx^2dy}h^2k + \frac{1}{2}\frac{d^3u}{dxdy^2}hk^2 + \&c.:$$

hence,

$$\frac{\text{solid } n'N'm'M' \ldots M}{hk} = \frac{d^2u}{dxdy} + \frac{1}{2}\frac{d^3u}{dx^2dy}h + \frac{1}{2}\frac{d^3u}{dxdy^2}k + \&c.$$

and passing to the limit, by making $h = 0$ and $k = 0$, the second member becomes $\frac{d^2u}{dxdy}$.

As regards the first member, the rectangle

$$n'N'm'M' = h \times k,$$

and the altitude of the solid becomes equal to $M'M = z$ when we pass to the limit: hence

$$\frac{d^2u}{dxdy} = z.$$

314. Although the differential coefficient

$$\frac{d^2u}{dxdy} = z.$$

has been determined by regarding u as a function of two variables, we can nevertheless return to the function u by the methods which have been explained for integrating a function of a single variable.

For we have

$$\frac{d^2u}{dx\,dy} = \frac{d\left(\frac{du}{dx}\right)}{dy} = z;$$

hence

$$\frac{d\left(\frac{du}{dx}\right)}{dy}dy = z\,dy;$$

and integrating under the supposition that x remains constant, and y varies, we have

$$\frac{du}{dx} = \int z dy + X',$$

whence

$$\frac{du}{dx} dx = dx \int z dy + X' dx;$$

and if we integrate this last expression under the supposition of x being the variable, and make $\int X' dx = X$,

$$u = \int dx \int z\, dy + X + Y.$$

It is plain that the constant, which is added to complete the first integral, may contain x in any manner whatever; and that which is added in the second integral, may contain y: *the first will disappear when we differentiate with respect to* y, *and the second when we differentiate with respect to* x.

The order of integration is not material. If we first integrate with respect to x, we can write

$$\frac{d^2 u}{dx\, dy} = \frac{d\left(\frac{du}{dy}\right)}{dx};$$

and by integrating, we find

$$\frac{du}{dy} = \int z\, dx, \quad u = \int dy \int z\, dx:$$

hence we may write

$$u = \int\int z\, dy\, dx, \quad \text{or} \quad u = \int\int z\, dx\, dy,$$

which indicates that there are two integrations to be performed, one with respect to x, and the other with respect to y.

315. If we consider the differentials as the indefinitely small increments of the variables on which they depend, we may regard the prism whose base is the rectangle $n'N'm'M'$, as composed of an indefinite number of small prisms, having equal bases, and a common altitude dz. Each one of these prisms will be expressed by $dx\,dy\,dz$, and we shall obtain their sum by integrating with respect to z between the limits $z=0$ and $z=MM'$, which will give

$$\int dx\,dy\,dz = z\,dx\,dy.$$

316. It is plain that $z\,dx$ is the differential of the area of the section made by the plane HQF parallel to the co-ordinate plane ZX; and consequently

$$\int z\,dx = \text{area of the section } HQF.$$

Hence, $(\int z\,dx)dy$ is equal to the elementary solid included between the parallel planes HQF, hqf, or

$$\int(\int z\,dx)dy = \int\int z\,dx\,dy$$

is equal to the solid which is limited by the surface and the three co-ordinate planes. If we consider a section of the solid parallel to the co-ordinate plane YZ, we have $\int z\,dy =$ area of the section EPG, and $\int\int z\,dx\,dy =$ solidity of the solid.

317. Let us suppose, as a first example, that

$$z = \frac{1}{x^2+y^2};$$

we shall then have

$$u = \int\int \frac{dx\,dy}{x^2+y^2} = \int dx \int \frac{dy}{x^2+y^2} = \int dy \int \frac{dx}{x^2+y^2}.$$

Let us now integrate under the supposition that x is constant; we then have

$$\int \frac{dy}{x^2+y^2} = \frac{1}{x}\text{tang}^{-1}\frac{y}{x} + X',$$

in which X' represents an arbitrary function of x. If we now make $\int X'dx = X$, and integrate again under the supposition that x is a variable, we have

$$\int dx \int \frac{dy}{x^2+y^2} = \int dx\left[\frac{1}{x}\text{tang}^{-1}\frac{y}{x} + X'\right]$$

$$= \int \frac{dx}{x}\text{tang}^{-1}\frac{y}{x} + X.$$

The integral of $\frac{dx}{x}\text{tang}^{-1}\frac{y}{x}$ is obtained in a series by substituting the value of (Art. 228),

$$\text{tang}^{-1}\frac{y}{x} = \frac{y}{x} - \frac{y^3}{3x^3} + \frac{y^5}{5x^5} - \frac{y^7}{7x^7} + \text{\&c.};$$

and since, in integrating with respect to x, we must add an arbitrary function of y, which we will represent by Y, we shall obtain

$$\int\int \frac{dxdy}{x^2+y^2} = X + Y - \frac{y}{x} + \frac{y^3}{9x^3} - \frac{y^5}{25x^5} + \frac{y^7}{49x^7} - \text{\&c.}$$

We shall obtain the same result by integrating in the inverse order, viz., by first supposing y to be constant. Under this supposition

$$\int \frac{dx}{x^2+y^2} = \frac{1}{y}\text{tang}^{-1}\frac{x}{y} + Y',$$

then integrating with respect to x,

$$\int dy \int \frac{dx}{x^2+y^2} = \int dy\left[\frac{1}{y}\text{tang}^{-1}\frac{x}{y} + Y'\right]$$

$$= \int \frac{dy}{y}\text{tang}^{-1}\frac{x}{y} + Y$$

But by observing that (Trig. Art. XVIII),

$$\text{tang}^{-1}\frac{x}{y} = \frac{\pi}{2} - \text{tang}^{-1}\frac{y}{x},$$

we shall have, after the second integration, and the addition of an arbitrary function of x,

$$\int\int \frac{dxdy}{x^2+y^2} = \frac{\pi}{2}\log y - \int \frac{dy}{y}\text{tang}\frac{y}{x} + Y + X;$$

and as we can include the term $\frac{\pi}{2}\log y$ in the arbitrary function Y, this result may be placed under the form

$$\int\int \frac{dxdy}{x^2+y^2} = X + Y - \int \frac{dy}{y}\text{tang}^{-1}\frac{y}{x},$$

which is the same as the result before obtained, as may be shown by placing for $\text{tang}^{-1}\frac{y}{x}$ its value, multiplying each term by $\frac{dy}{y}$, and integrating.

318. When we consider

$$\int\int zdxdy$$

as expressing the solidity of a solid, it is necessary to consider the limits between which each integral is taken, and these limits will depend on the nature of the solid whose cubature is to be determined. Let it be equired, for example, to find the solidity of a sphere, of which the centre is at the origin of co-ordinates. Designating the radius by R, we have

$$x^2+y^2+z^2=R^2,$$

and consequently,

$$\int\int z dx dy=\int\int dx dy\sqrt{R^2-x^2-y^2}.$$

If now, we suppose y constant, and make $R^2-y^2=R'^2$, and then integrate with respect to x, we have

$$\int dx\sqrt{R^2-x^2-y^2}=\int dx\sqrt{R'^2-x^2},$$

and integrating this last expression, first by formula (B) Art. 239, and then by Art. 220, we have

$$\int dx\sqrt{R'^2-x^2}=\frac{x}{2}\sqrt{R'^2-x^2}+\frac{1}{2}R'^2\sin^{-1}\frac{x}{R'}+Y;$$

and substituting for R'^2 its value, we obtain

$$\int dx\sqrt{R^2-x^2-y^2}=\frac{x}{2}\sqrt{R^2-x^2-y^2}+\frac{1}{2}(R^2-y^2)\sin^{-1}\left(\frac{x}{\sqrt{R^2-y^2}}\right)+Y.$$

It should be remarked, that $\int z dx$ expresses the area of a section of the sphere parallel to the co-ordinate plane ZX, for any ordinate $y=AQ$, and to obtain this area we must integrate between the limits $x=0$ and $x=QF$. But since the point F is in the co-ordinate plane YX, we have for this point $z=0$, and the equation of the surface gives

$$QF=x=\sqrt{R^2-y^2};$$

therefore, for every value of y the integral $\int z dx$ must be taken between the limits $x=0$ and $x=\sqrt{R^2-y^2}$. Integrating between these limits we have

$$\int dx\sqrt{R^2-x^2-y^2}=\frac{1}{2}(R^2-y^2)\sin^{-1}(1)$$

$$=\frac{\pi}{4}(R^2-y^2),$$

since, $\sin^{-1}(1)=\frac{\pi}{2}$:

hence,

$$\int dy\int zdx=\frac{\pi}{4}\int dy(R^2-y^2)=\frac{\pi}{4}\left(R^2y-\frac{y^3}{3}\right)+X,$$

and taking this last integral between the limits $y=0$ and $y=AC=R$, we obtain

$$\frac{\pi R^3}{6},$$

which represents that part of the sphere that is contained in the first angle of the co-ordinate planes, or one-eighth of the entire solidity. Hence,

$$\text{solidity of the sphere }=\frac{4}{3}R^3\pi=\frac{1}{6}D^3\pi.$$

We might at once find the solidity of the hemisphere which is above the horizontal plane YX, by integrating between the limits

$$x=-\sqrt{R^2-y^2}\quad\text{and}\quad x=+\sqrt{R^2-y^2}.$$

Taking the integral between the limits

$$x=0\quad\text{and}\quad x=-\sqrt{R^2-y^2},$$

we have $\int zdx=-\frac{\pi}{4}(R^2-y^2)$;

and between the limits

$$x=0\quad\text{and}\quad x=+\sqrt{R^2-y^2},$$

we have $$\int z dx = \frac{\pi}{4}(R^2 - y^2);$$

hence, between the extreme limits, we have

$$\int z dx = \frac{\pi}{2}(R^2 - y^2).$$

Then taking the integral

$$\int dy \int z dz = \frac{\pi}{2} \int dy (R^2 - y^2)$$

between the limits

$$y = -R \quad \text{and} \quad y = +R,$$

we find the solidity to be

$$\frac{2}{3} R^3 \pi;$$

or the solidity of the entire sphere is,

$$\frac{4}{3} R^3 \pi.$$

THE END

www.ingramcontent.com/pod-product-compliance
Lightning Source LLC
LaVergne TN
LVHW010226110826
845151LV00004B/1193

* 9 7 8 1 4 2 5 5 2 6 7 6 4 *